Infectious and Medical Waste Management

Peter A. Reinhardt
Judith G. Gordon

LEWIS PUBLISHERS

Library of Congress Cataloging-in-Publication Data

Reinhardt, Peter A.
    Infectious and medical waste management / by Peter A. Reinhardt
and Judith G. Gordon
        p.   cm.
    Includes bibliographical references.
    Includes index.
    1. Infectious wastes—Handbooks, manuals, etc. 2. Medical wastes—
Handbooks, manuals, etc. 3. Medical centers—Waste disposal—
Handbooks, manuals, etc. I. Gordon, Judith G. II. Title.
    [DNLM: 1. Hazardous Waste—prevention & control. 2. Refuse
Disposal—methods. 3. Waste Products. WA 778 R371i]
    RA567.7.R45   1991   363.72'88—dc20   90-6193
    DNLM/DLC   CIP
    for Library of Congress
    ISBN 0-87371-158-0

LEWIS PUBLISHERS, INC.
121 South Main Street, Chelsea, Michigan 48118

PRINTED IN THE UNITED STATES OF AMERICA

Preface

This book is intended for use by those professionals who are responsible for the planning and implementation of infectious and medical waste management programs. It will help in decisions on how best to manage infectious and medical wastes and will also be of interest to those who are concerned about safe waste management. The authors especially hope that this book will assist health service administrators in understanding their responsibilities for protecting their employees, patients, visitors, the community, and the environment from the risks inherent in the management and disposal of infectious and medical wastes.

The authors' goal has been to present useful information that will help to reduce these risks. However, because of the broad scope of this book, differing local and state laws and regulations, and the many changes occurring in the field of infectious and medical waste management, the information presented in this book cannot be complete and up-to-date at any given moment. It is the responsibility of the reader to stay informed about changing regulations and liabilities.

The authors believe that the information and data included in this book are true and accurate. Recommendations and suggestions are made without warranty or guarantee of any kind. Consequently, the authors assume no responsibility connected with the use of this book or information included herein. The authors shall not be responsible for any incidental or consequential damages whatsoever.

Acknowledgment

We thank Harvey W. Rogers of the Agency for Toxic Substances and Disease Registry (ATSDR), and formerly of the Division of Safety at the National Institutes of Health, for reviewing our manuscript. His constructive comments, helpful suggestions, and enthusiastic support were invaluable.

Peter A. Reinhardt is the Hazardous Waste Officer and Chemical Safety Supervisor of the University of Wisconsin at Madison. In that capacity he oversees the operation of the University's eight pathological incinerators, waste collection program, and two hazardous waste storage facilities. He has been responsible for hazardous waste management at the University of Wisconsin–Madison since 1980. Mr. Reinhardt also is an instructor in the Department of Professional Engineering Development, teaching a course on infectious and medical waste management. He lectures on waste management concepts, steam sterilization of infectious waste, and responding to hazardous materials emergencies.

He is an appointed member of the National Committee on Clinical Laboratory Standards' subcommittee on waste management and the American Chemical Society's task force on hazardous waste. He has written several articles on waste management as well as several chapters of *Hazardous Waste Management at Educational Institutions* (National Association of College and University Business Officers, 1987).

Mr. Reinhardt has a BS in biochemistry from the University of Wisconsin–Madison.

Judith G. Gordon is President of Gordon Resources Consultants, Inc., of Reston, Virginia, a company that provides consulting services on waste management and regulatory compliance to hospitals and industry. A specialist in medical waste management and control, Mrs. Gordon evaluates practices for the management of infectious and hazardous wastes, develops systems for waste management, prepares implementation plans, analyzes regulatory requirements, and provides training programs. Her clients include a variety of private and governmental organizations. She has been an expert consultant to the United States Environmental Protection Agency on the Resource Conservation and Recovery Act hazardous waste program.

Mrs. Gordon is a frequent speaker at seminars and workshops on the management of infectious and hazardous wastes, on federal and state regulations, and on occupational hazards for waste handlers. She participates as a course instructor at the University of Wisconsin–Madison through the Department of Engineering Professional Development.

She is a charter member of the American Biological Safety Association, for which she has served as a member of the Executive Council, as Chairperson of the Constitution and Bylaws Committee, and as a representative on the American Society for Testing and Materials Subcommittee on Medical Waste. She is also a member of the Association for Practitioners in Infection Control and the American Society for Microbiology.

Mrs. Gordon's publications include the Environmental Protection Agency's Draft Manual for Infectious Waste Management and other writings on waste management and the health effects of occupational and environmental exposures. She is co-editor of *Biohazards Management Handbook.*

She received a BA from the University of Pennsylvania and an MA from the University of California at Berkeley.

Contents

SECTION I
INTRODUCTION

1. The Search for a Successful Waste Management Strategy ...   3

2. Regulations and Standards for Infectious and Medical Waste
   Management.......................................   19

SECTION II
INFECTIOUS WASTE MANAGEMENT

3. Identification of Infectious Waste   31

4. Handling, Storage, and Transport of Infectious Waste   43

5. Treatment Considerations and Options   55

6. Steam Sterilization of Infectious Waste...................   71

7. Incineration of Infectious Waste   93

8. Other Treatment Technologies..........................   109

9. Disposal of Treated Waste   125

10. Minimizing Infectious Waste   129

SECTION III
MANAGEMENT OF OTHER MEDICAL WASTE

11. Antineoplastic Drugs and Other Chemical Wastes..........   139

12. Managing Low-Level Radioactive Waste   159

13. Wastes with Multiple Hazards   177

SECTION IV
KEEPING YOUR SYSTEM GOING

14. Occupational Safety for Waste Management 185

15. Preparing for Hazardous Material Emergencies 197

16. Training Staff and Waste Handlers 221

17. Completing the Process: Essential Components of Effective
    Waste Management 231

APPENDICES

Appendix A   A Guide to the Medical Waste Tracking
             Regulations 245

Appendix B   U.S. EPA and State Hazardous Waste Contacts ... 259

Appendix C   Infectious Waste Management Audit 265

Index 269

Infectious and Medical Waste Management

SECTION I

Introduction

CHAPTER 1

The Search for a Successful
Waste Management Strategy

INTRODUCTION

Institutions generating infectious and medical waste have found its management to be an intractable problem. New rules and guidelines are continually being issued by various regulatory bodies. Employees complain of threats to their health and inadequate training. Commercial services for infectious and medical waste disposal are either poor or nonexistent in most areas of the country. Even states with strict laws governing infectious waste lack clear guidance. Few inspectors understand the nature of its generation or the labyrinth of its control. Thus, institutional administrators fail to find a waste management system that is both workable and cost-effective. And loose hypodermic needles still end up at the local landfill.

The need for infectious and medical waste management now reaches beyond hospitals and medical centers to smaller waste generators, such as clinics, colleges and universities, diagnostic laboratories, pharmaceutical and biotechnology companies, funeral homes, vocational/technical schools, doctor's offices, and other health service facilities. An administrator looking for a solution faces a perplexity of jargon, management options, and environmental nuances, including such controversial issues as employee safety and acquired immune deficiency syndrome (AIDS).

This book should make sense of it all. The business officer, institutional attorney, and risk manager of a medical facility will gain a perspective on the risks of medical wastes and the purpose of waste management. Critical details of how waste should be managed are provided for the biosafety officer, engineer, infection control staff, and environmental services supervisor.

This chapter serves as an overview of waste management concepts, the risks posed by infectious and medical wastes, and the objectives of waste management. As described in this chapter, management means active controls, and that requires a written plan and administrative support.

Chapter 2 describes federal and state laws and national standards for infectious and medical waste. Chapters 3 through 10 explain the management methods typically used to manage infectious and medical waste. Because medical facilities also generate chemical and low-level radioactive wastes, as well as wastes with multiple hazards, Chapters 11, 12, and 13 make this book a complete guide to these problematic wastes. A management program for both hazardous *material* and hazardous waste must include occupational safety, contingency planning, and employee training; Chapters 14, 15 and 16 deal with these subjects. Lastly, Chapter 17 covers the remaining administrative components that are essential for a complete management program.

Although the authors stress the importance of an integrated waste management plan, each chapter stands well on its own. No one book can provide all the information needed in this area, so references to key resources are listed throughout.

UNDERSTANDING WASTE AND WASTE MANAGEMENT

A quick review of the jargon used in waste management is a good place to begin. *Solid waste* is a catch-all term used by the U.S. Environmental Protection Agency (EPA) to define all solid, liquid, and gaseous waste.[1] Chemical, hazardous, infectious, and medical wastes are subcategories of solid waste that can threaten human health or the environment because they are potentially harmful. The term "solid waste" is frequently used generically for the nonhazardous component of solid waste, such as normal refuse and trash. Unless otherwise noted, the authors use the term "solid waste" in that nonlegal sense, to denote normal trash that doesn't have any inherently harmful characteristics that merit additional regulation. The topic of this book is the need to take special precautions (relative to other solid waste) with certain chemical, radioactive, infectious, and other medical wastes that have the potential to harm human health or the environment.

Hazardous waste is legally defined by EPA in Title 40 of the *Code of Federal Regulations* (40 CFR), Part 261. EPA's use of the term "hazardous waste" is very confusing because it only pertains to hazardous *chemical* waste. Although infectious waste is indeed hazardous, it doesn't currently fit EPA's legal definition of hazardous waste. (See Chapter 11 for a detailed discussion of hazardous chemical waste.) Also confusing is the term "hazardous material," which is defined by the U.S. Department of Transportation (DOT) to include chemical, radioactive, and etiological agents.[2] EPA also uses the term "hazardous substance" in some environmental laws. Consequently, the word "hazardous" must be used carefully. In this book, it is used in its nonlegal generic sense unless otherwise noted.

Infectious and medical waste management is the primary focus of this book. In 1976, Congress used the word "infectious" to characterize a potential type of hazardous waste. Until 1988, EPA continued to use the word in its guidance for "waste capable of producing an infectious disease."[3] EPA now uses the term "medical waste," which includes many of the wastes formally listed as infectious. Chapter 2 discusses the definitions of and differences between infectious waste and medical waste as they are used by regulatory authorities. Chapter 3 describes the various types of infectious and medical waste.

Generation is the term used for the process of making waste. For example, the care of patients often generates infectious waste. Waste generation also refers to the decision to dispose of a material, such as the discarding of laboratory chemicals prompted by an annual review of storage shelves.

At some point, a *determination* must be made as to whether a waste merits special handling and disposal precautions. For many types of chemical, radioactive, infectious, and medical waste, special procedures to make this determination (e.g., identification, tests) are required by law. Institutions are encouraged to also evaluate wastes that are not regulated but have risks that may warrant additional precautions.

The determination can be made prior to actual waste generation if it can be based on the waste's identity, knowledge of the material or process from which the waste is generated, or lists based on such identification or knowledge. If such information is not available at the time a waste is generated, tests are needed to measure a hazardous characteristic. Amount and concentration can affect the determination.

There are many categories and subcategories of hazardous wastes. The term *waste stream* is used to distinguish a segregated waste type; sharps and flammable solvents are examples of two such waste streams.

It is from this point onward that *containment* is important to prevent contamination or exposure to workers handling the waste. Wastes are often temporarily *accumulated* at the point of generation. *Storage* usually refers to the use of a dedicated facility or centralized area where the generated waste is held prior to transport, treatment, or disposal.

Treatment refers to the process that reduces or eliminates the hazardous characteristics or reduces the amount of a waste. *Disposal* usually refers to the permanent containment of a waste in a landfill. (The waste may or may not retain some harmful characteristics.) Disposal sometimes denotes dilution and dispersal, such as emission into air of small amounts of contaminants from incineration or discharge into sewers.

BASIC WASTE MANAGEMENT CONCEPTS

Surprisingly, the themes of waste management are similar, whether addressing infectious, chemical, or radioactive waste, or even normal trash. Three themes are prominent:

- Waste should be *managed.* Too often waste disposal is viewed as an isolated problem, such as the decision as to which receptacle to use for a handful of waste, or a full trash can waiting to be emptied, or an autoclave in need of maintenance. Today, however, disposal is viewed as only one part of a larger waste management system. Waste is now *managed* through a pathway that also includes generation, segregation, collection, storage, processing, transport, and treatment. (See Figure 1.1.) Each step carries with it its own risks and costs. Thus, management requires analysis and active control from generation through disposal.

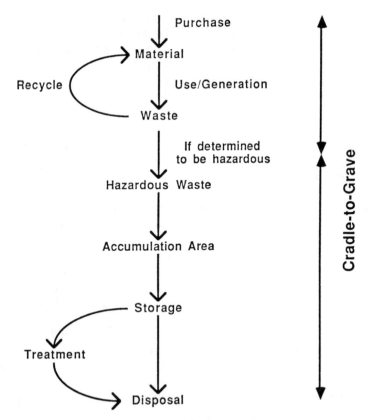

Figure 1.1 Hazardous waste management: basic concepts.

- Where possible, hazardous waste should be *treated* prior to disposal to reduce or eliminate its hazard. (Because characteristics differ for each waste type, all wastes cannot be dealt with in the same way.)
- *Waste minimization* or reduction is undoubtedly the most desirable goal of waste management. Waste minimization (described in more detail in Chapters 10, 11, and 12) may be brought about by material substitution, waste segregation, recycling, reuse, procedural changes, acquisition constraints, and using treatments and processes that reduce a waste's amount or hazard. It is often prudent to extend the scope of waste management to monitoring the use of materials before they become wastes with an eye to reducing the amount of waste by modifying waste-generating activities.

RISKS OF INFECTIOUS AND MEDICAL WASTES

It is critical that institutional administrators and waste managers incorporate complete understanding of the risks of infectious and medical wastes into their institution's management plan. As makers of institutional policy, administrators are responsible for providing a safe workplace and preventing environmental damage. If their institution is deemed negligent, they may be held personally liable due to their decisionmaking responsibilities.

The last two decades have been accompanied by tremendous changes in the scope of environmental and occupational liability. It is prudent for administrators to regularly consult their institution's attorney, risk manager, safety director, biosafety officer, hazardous materials manager, and infection control staff to keep up to date. They should encourage these professionals to stay abreast of developments through continuing education and review of the literature.

Risk is measured as the product of the consequence of an event and its probability. For waste, consequence can take the form of workers' compensation claims, fines, lawsuits, loss of accreditation, or even criminal penalties for executives. Risks of waste can generally be classified as occupational, environmental, legal, political, social, or economic. Sometimes there are multiple, related risks, such as the burning of regulated chemical waste in a municipal incinerator: it is not only illegal, but may result in an explosion and the production of toxic ash or offgases.

Occupational Risks

Occupational risk has long been identified as the most serious hazard of infectious and medical waste. Potential occupational exposures include direct exposure to patients, visitors, and workers who handle or come into proximity to waste.

Employees who are injured in the course of their work with waste have the right to be compensated. In some instances of workplace injury, the injured person may sue an employer for negligence without workers' compensation limits.[4]

Occupational risks are also a serious concern outside of the institution. Normal trash is subjected to a great deal of scrutiny in many parts of the country. Landfill and refuse incinerator operators attempt to segregate wastes for recycling and other disposal routes (e.g., newspapers, tires, metals, plastics, etc.). Municipal waste facilities are more likely to carefully screen waste from businesses and institutions to prevent entry of those wastes that are perceived to be hazardous to their workers or the environment. These activities greatly increase the chances of occupational exposure if infectious waste is left untreated. Needle sticks (puncture wounds from hypodermic needles or other sharps) are a major concern of municipal waste workers.

Environmental Risks

Environmental risks include the possibility of a release of waste to groundwater, surface water, or air. The uncontained disposition of waste (such as open dumping) also creates an environmental hazard to the community because of the potential for injury or disease transmission by direct contact.

Waste treatment methods carry significant environmental risks. As described in Chapter 6, using steam sterilization (autoclaving) necessitates special precautions to prevent release of infectious agents to the air via the steam exhaust or to the sewer via the drain pipe. This is also true of incinerators, especially those built before 1970 or those that are poorly run. As discussed in Chapter 7, incinerators may not destroy infectious agents as expected, releasing them instead to the air, in the ash, or via scrubber effluent. Toxic chemicals produced from combustion, which may be emitted from the stack or concentrated in the ash, may be another serious problem with waste incineration.

The environmental risks of disposal practices are well known. Even small amounts of laboratory solvents, when disposed of in a landfill, can leach into drinking water. Some infectious waste can stay infectious for many years if disposed of without being sterilized although this is uncommon. For example, anthrax-infected cattle disposed of by burial contain spores that are known to survive for many decades in dry soil.

Legal Liabilities

The scope of legal liability associated with waste management is enormous, and includes both occupational and environmental laws. The extent of exposure, breadth of litigation, quagmire of laws, and continuing developments will amaze even an experienced environmental manager. It is nearly impossible for a waste generator to avoid these potential liabilities.

Laws specify that, under many conditions, the generator is liable for any harm caused by its waste from the point of generation through every step in the pathway, including disposal. This "cradle-to-grave" liability applies regardless of the generator's control over the management methods employed. If a contractor handles or disposes of waste improperly, it nevertheless may be the responsibility of the generator to pay for remedial actions, including disposal site cleanup, and compensation to injured parties.

Operators of commercial waste disposal services (transporters, incinerators, landfills, etc.) face enormous liabilities also, which are often uninsurable. Every organization or company associated with the waste is likely to share some responsibility if occupational or environmental damage occurs.

Although it is possible to share the environmental and occupational risks of waste management with the transporter, the commercial disposal service, and other generators, federal and state laws also stipulate that in many cases, an institution is forever liable for its waste and the harm it may cause. For example, a state-of-the-art, legally approved treatment method of today may be found a decade from now to have created an environmental hazard at a disposal site. The generators that used the site will be responsible for remediation, and if other users of the site (or the site operator) are no longer in business, the share of cleanup costs due from generators still existing will be disproportionately higher than the amount of waste they contributed.

Legal liability can take many forms, including:

- Noncompliance with federal, state, and local laws. Day-to-day compliance with occupational and environmental laws is very difficult, given their number, complexity, and frequent changes. Noncompliance can result in compliance orders, consent decrees, citations, fines, criminal charges, and other penalties.
- Infraction of standards set by accreditation and certifying bodies. As described in Chapter 2, an accrediting body (e.g., Joint Commission on the Accreditation of Healthcare Organizations [JCAHO]) may cite an institution for failure to meet standards far less stringent than set by law.
- Lawsuits, which are easily provoked in the event of an inadvertent exposure, an accidental release, or a spill. All that is needed is the perception of environmental or occupational injury. Also, an institution found to be in violation of waste standards or laws is open to lawsuits that base their claims on those findings.
- Citizen suits, which are increasingly included in environmental and occupational safety laws. They allow any person or organization to sue an institution for noncompliance without regard to findings by a government agency.

Remember, though, that legal liability is only one risk of infectious and medical waste. Remember also that compliance with current waste management laws *does not* eliminate occupational and environmental risks or prevent future legal liabilities.

Political, Economic, and Social Risks

Many of the risks of infectious and medical waste are not obvious. Science traditionally measures consequence by analysis of morbidity and mortality statistics. But even if the risk measured by these methods is very small, the public's perception of risk may be as valid, more important, and a more accurate estimation of risk.[5] The many hours spent on public relations if untreated wastes are found at the local landfill, or the time necessary to negotiate zoning changes to construct an incinerator, are costs (consequences) just as real as environmental damage.

It is important to appreciate how the public perceives risk. Measurable incidence of disease and environmental harm may be not as important as appearances, property values, whether the environmental impact is shared equally within a community, or whether there is a mechanism for public/governmental control. As stated by Milton Russell, an economist at Oak Ridge National Laboratory and the University of Tennessee, "It is simplistic to believe that people have only one goal in protecting the environment — to reduce the calculated risk. They are also concerned about physical characteristics of the risk, its source, how it is distributed, and whether it is fairly imposed upon them."[6] Although risk management requires evaluation of the probability of a harmful event, society and policymakers have increasingly focused on the potential harm a hazardous material may cause, however improbable, occurrence of that harm may be.

Without controls and a plan, institutions will likely pay higher costs for waste disposal as a result of employee and service disruptions. Employees and their unions have identified infectious and medical waste as a serious workplace hazard. Disposal firms and facilities can refuse to serve a careless institution (e.g., from which infectious waste and normal trash are not kept separate), and the threat of liability may obligate them to do so. Environmental, safety, and space concerns have severely stressed the nation's solid waste disposal business. As a result, operators do not hesitate to restrict, and sometimes deny, access to their landfills if they perceive improper waste management. Medical and health care facilities have been singled out as a particular liability.

Although the hazards of infectious and medical waste have long justified its careful management, associated risks (particularly legal and political risks) have increased sharply in recent years. The dramatic change in managing everyday trash in the last decade exemplifies this increase.

NATIONAL WASTE MANAGEMENT TRENDS

There is a solid waste crisis in the United States caused by increasing amounts of waste and decreasing capacity of landfills, the disposal route most often used.[7]

To relieve the problem, some communities have built trash incinerators that greatly reduce waste volume. However, the general public is very concerned with the degradation of our environment from whatever source. Thus, even when the environmental impact is very small, there is tremendous public opposition to constructing both incinerators and landfills. Local approvals are usually required – and denied. Recycling and waste minimization appear to be publicly acceptable, although their costs may be considerably higher. Waste minimization is expected to be a major theme of the reauthorization of the Resource Conservation and Recovery Act (RCRA), scheduled for 1990 or 1991.

Infectious and medical waste is very much caught in the middle of this crisis. Because of efforts to control costs, the danger of AIDS, and an increased concern for occupational safety, medical institutions are using more disposable items and the volume of infectious and medical wastes has increased.[8] Another reason for the increasing volume is the more stringent laws, defining infectious and medical waste broadly, that many states have recently adopted; items that previously could be disposed of as normal trash must now be sterilized.

Even before the 1988 beach washups, states were increasingly reviewing infectious and medical waste management activities.[9] Some state air quality departments have responded to citizen concerns about incineration by restricting and closing (or planning to close through strict regulation) hospital incinerators. (The Clean Air Act is scheduled for reauthorization in 1990 and may contain new federal laws for hospital incineration.) Because of these new regulations, many medical institutions must now employ offsite commercial disposal services.

These specialized services and higher volume mean sharply increased costs for the disposal of infectious and medical waste. These trends will continue as the Medical Waste Tracking Act (MWTA) of 1988 is implemented. (The MWTA is described in Chapter 2; it also defines medical waste broadly.)

LESSONS FROM THE 1988 BEACH WASHUPS

In the summer of 1988, syringes, IV tubing, and prescription bottles (along with a lot of other nonmedical refuse) washed up on the shores of five east coast states.[10] As a result, beaches were closed, tourists were frightened away, and beach-related businesses may have lost $1 billion. When reporters investigated, they found a solid waste disposal system out of control, with few measures to prevent ocean dumping. They also found that infectious and potentially infectious wastes from medical institutions were largely unregulated. In many cases, there was either no law or no enforcement to prevent the disposal of unsterilized infectious waste in the normal trash.

No one was able to determine if the syringes had been sterilized or not, but all medical professionals agreed that used syringes should not be disposed of in the ocean or on a beach. To the public, medical waste on the beaches meant that medical institutions don't care enough about the environment and the disposal of their waste; new controls were necessary to prevent a recurrence.

The Congressional Office of Technology Assessment estimates that more than 3.2 million tons of infectious and medical waste are generated every year.[8] Of that, only 3 cubic yards were found on the beaches in 1988. In many cases, the wastes were suspected to originate from illicit drug use or at-home medication, sources that are nearly impossible to control.

There is a great deal of debate over the issue of infectious and medical waste regulation. Many medical professionals believe such regulation is unnecessary because current practices (many of them self- or industry-imposed) sufficiently protect human health and the environment from the small risk posed by these wastes. It appears, however, that as science is able to measure ever smaller risks, the public's tolerance of risks becomes smaller, too. A preventable risk has become an unacceptable risk. As demonstrated by the political and economic cost of beach washups, public perception of risk may as well be considered real risk. Risk communication (the effort to improve the understanding of risk) may be able to change perceptions, but it is a very difficult task.[11]

OBJECTIVES OF WASTE MANAGEMENT

Institutions can exert considerable control over the management of infectious, chemical, radioactive, and other medical waste, unlike normal refuse. This first requires finding and committing to a strategy for waste management, then adopting a system that is well supported, flexible, and thorough.

A Need for Strategy

Waste management deserves analysis, planning, and cost containment, just as do other major building support services. Without a strategy, liabilities grow without restraint and every breakdown of equipment or procedures is a crisis. From the previous discussion of risks, four strategic objectives can be identified:

1. *Reduce risks and liabilities.* The remaining chapters of this book explain ways to reduce the risks of waste management. Only executive-level decisions, however, enable an institution to comprehensively control wastes and their risks, particularly those decisions regarding budget and staffing. Decisions to reduce and manage risks should be documented in policies, further detailed in employee procedures, and implemented through training.
2. *Control costs.* Because waste disposal has grown to be a significant institutional support cost, waste management deserves careful evaluation. Chapter 17 explains how to conduct an audit of current practices while exploring opportunities for waste minimization.

3. *Plan for the future.* Much is unknown about the future of waste regulation, available services, and practices that generate wastes. Lacking assurances with respect to the future, business has found that it is best to diversify management plans. Diversification can be achieved by using a variety of treatment technologies and both onsite and offsite facilities (e.g., an institutional steam sterilizer and a commercial incinerator). There is some safety in selecting technologies in wide use that have a proven record. Self-reliance in treatment methods (e.g., institutional ownership of an incinerator) insulates the institution from disruptions in commercial disposal services. Joining with other generators in cooperative facilities and contracts, a topic of Chapter 17, also helps diversify solutions.

Contingency plans are also important to protect against breakdowns and future changes. Planning for hazardous material emergencies is the focus of Chapter 15 and coping with other failures in the waste management system is discussed in Chapter 17.

4. *Communicate the institution's commitment to protecting human health and the environment.* Averting political risks can be as simple as listening to employees and their union representatives, including a public representative on the safety committee, and discussing waste management challenges before civic groups. Such openness, however, is risky in itself; sometimes it backfires, resulting in the community focusing on a dormant problem. Still, low-profile risk communication nearly always pays off. It helps ensure that, should a waste crisis occur, there are members of the community that can put it in perspective with the institution's positive efforts.

No one strategy will meet the needs of all institutions, nor all the needs of any single institution. Instead, administrators need to review the problem, carefully evaluate the options presented here, and select those that are most appropriate for their wastes and institution.

PLANNING FOR WASTE MANAGEMENT

A written waste management plan is tangible evidence of a serious commitment to safely manage infectious and hazardous waste. The plan should outline the current procedures as well as proposals for improving waste management. Although the plan can touch on solid waste disposal, it is best for the focus to be on infectious and medical waste management—the most problematic areas. A sample outline is given in Table 1.1. Also, flow charts are useful for illustrating complex waste management operations; see Figure 1.1 for an example.

The plan should be prepared by the person or committee responsible for waste management oversight. It must identify functional responsibilities for all

Table 1.1.   Sample Institutional Waste Management Plan Outline

I. Policy and Purpose of the Waste Management Program

II. Scope of Management Plan
   A. Waste types
   B. Activities that generate infectious and medical wastes

III. Current Management Methods
   A. Identification
   B. Collection
   C. Storage
   D. Treatment
   E. Transport
   F. Disposal

IV. Individual Responsibilities and Employee Training

V. Waste Minimization Efforts

VI. Occupational Safety

VII. Emergency Response

VIII. Quality Assurance

IX. Annual (or Biennial) Reports
   A. Assessment of current management methods
   B. Costs of waste management
   C. Legal and other liabilities
   D. Five-year planning
   E. Operational needs

procedures. When the plan is completed, have it approved by the organization's administrators responsible for planning. An annual update of the plan is best (and is required by the Joint Commission for the Accreditation of Healthcare Organizations).

Getting Additional Input

The evaluation of current practices should consider the opinions of various staff. Although custodial staff and waste handlers may lack technical expertise, their perspective from hands-on experience in waste handling is invaluable. Interviews with operators of current disposal sites and firms and with state and local regulators can reveal trends in infectious waste management.

It is wise to seek the advice of institutional experts who can assess hazards, liabilities, and proposals. Planning and decisions should include the organization's attorney, risk manager, environmental services supervisor, infection control officer, safety director, physical plant director, business officer. An outside consultant can provide specialized expertise. Because oversight of infectious and medical waste management requires guidance from a variety of perspectives, the planning (and later the oversight) of management activities is commonly done by a committee of similar composition.

Preparation of a plan is particularly important for larger organizations, as they typically deal with changes in building services through written plans and proposals. Many large institutions regularly prepare five- or ten-year plans that list needed capital improvements and other expenditures. Waste management—its needed staffing, supplies, and facilities—should be included in that planning.

Decisionmaking

Key to the success of any management program is the involvement and support of administrative personnel—institutional decisionmakers. Willing or not, institutional executives are involved in waste management. Regulators and the courts often place the responsibility for safety on the executives who make the decisions—including the decision to ignore a problem.

Budgeting

The system must be sufficiently supported by a budget that allows not only for proper collection and disposal, but also for adequate staff training. Institutional executives have the authority to reallocate resources to accomplish the plan's objectives; consequently, they must understand the need for such planning.

IMPLEMENTING THE PLAN

An outline of an institutional waste management plan is given in Table 1.1. Perhaps the three most important components of the plan are its policies, procedures, and assignment of responsibilities.

Policies

Any institution or company that generates infectious or medical wastes should have a formal policy that defines the objectives for managing the risks of these wastes. The policy should be affirmed by the organization's chief executive officer. Ideally, the objectives should be set forth as part of a comprehensive policy that addresses risks of any hazardous materials or wastes. The policy should set overall goals of preserving the health of employees, patients, visitors, and others who may encounter hazardous waste; protecting the environment and the community; and ensuring regulatory compliance.

Procedures and Training

Written procedures are needed to explain how the policy objectives should be accomplished. A formal procedure manual is essential. It is an important reference for institutional personnel. Effective training of employees utilizes a variety of written, graphic, and interactive training methods, including orientation presentations, one-on-one training by supervisors, signs and other written notices, routine updates for supervisors, and regular refresher training. Written documentation—showing that procedures have been followed and employees have been adequately trained—is important for limiting an institution's liability.

Assigning Responsibilities

The procedures need to specify responsibilities and duties by position. Although service personnel have traditionally been responsible for infectious and medical waste management, it requires oversight by a person with a broader knowledge of regulations and risk management. Some institutions designate a hazardous waste or medical waste manager, responsible for supervising all aspects of the waste management system. In any case, many employees have a hand in waste management; thus, administrative support, coordination of assignments, and communication are essential.

A FLEXIBLE STRATEGY FOR TOMORROW

Today waste is seen as a serious threat to our land, water, and air resources. Sound procedures for waste management seek to minimize damage to these media. Institutional administrators are responsible for understanding how waste is managed, how it is being disposed of, and what its risks are.

Beyond careful management, institutions must gain the trust of their employees and the community. Tangible signs of responsible hazardous materials management include well-defined objectives, an effective training program, visible administrative support, and participation in community planning for hazardous materials emergencies. When these are in place, it may be a good time to suggest a story on the challenges of waste management to the local newspaper editor.

Changes in waste management strategies will likely be necessary as more is learned about risks, as regulators refine controls and as society's perceptions of what is safe are refocused. The concepts discussed here should provide a firm foundation on which to base decisions in the face of such challenges.

REFERENCES

1. *Code of Federal Regulations* Title 40: Part 240, Section 240.101(y).
2. *Code of Federal Regulations* Title 49: Part 171, Section 171.8.
3. U.S. Environmental Protection Agency. "EPA Guide for Infectious Waste Management." EPA/530-SW-86-014. U.S. EPA, Office of Solid Waste and Emergency Response, Washington, DC (May 1986).

4. James, A. N. "Legal Realities and Practical Applications in Laboratory Safety Management," in J. H. Richardson et al., Eds. *Proceedings of the* 1985 *Institute on Critical Issues in Health Laboratory Practice: Safety Management in the Public Health Laboratory* (Wilmington, DE: Du Pont Co., 1985), pp. 177-182.

5. Freudenburg, W. R. "Perceived Risk, Real Risk: Social Science and the Art of Probabilistic Risk Assessment." *Science* 242:44–49 (1988).

6. Russell, M. "Risk Communication: Informing Public Opinion." *EPA Journal* 13(9):20 (1987).

7. O'Leary, P. R., P. W. Walch, and R. K. Ham. "Managing Solid Waste." *Sci. Am.* 259(6):36–42 (1988).

8. "Issues in Medical Waste Management – Background Paper." U.S. Congress, Office of Technology Assessment, OTA-BP-O-49, U.S. Government Printing Office (October 1988), p. 1.

9. Gellhaus, K. L. and B. C. Melewski. "Hemorrhage from the Hospitals: Mismanagement of Infectious Waste in New York State." New York State Legislative Commission of Solid Waste Management (March 25, 1986).

10. Lee, M. R. "Infectious Waste and Beach Closings." Congressional Research Service Report 88-596 ENR (September 9, 1988).

11. Sandman, P. M. "Explaining Environmental Risk." U.S. EPA Office of Toxic Substances (November 1986).

CHAPTER 2

Regulations and Standards for Infectious and Medical Waste Management

Various regulations, guidelines, and standards affect the management of infectious and medical wastes. *Regulations* are issued by governments at the federal, state, and local levels. Regulatory requirements are mandatory and enforceable by law, and penalties can be assessed for noncompliance. In contrast, *guidelines* are recommendations; following guidelines is therefore voluntary rather than mandatory. Guidelines are usually issued by government agencies and by professional organizations. *Standards* are of two types (for performance of activities and for quality of products); they are usually set by professional organizations. Standards are not enforceable per se, although some types of certification are contingent on the meeting of certain standards.

Only regulatory requirements are mandatory, whereas standards and guidelines are usually voluntary. However, regulatory agencies sometimes incorporate standards and guidelines into the regulations. If this is done, these standards and guidelines then become mandatory and enforceable. Another consideration regarding standards and guidelines is how they are regarded by the community; for a discussion of their impact and importance, see the section entitled ''Community Standards'' in this chapter.

REGULATIONS

Regulations are promulgated by government regulatory agencies at the federal, state, and local levels. Regulations that affect the management of infectious and medical wastes can be promulgated at any level of government. So far, such action has not been taken in every jurisdiction in the country, and the regulatory requirements vary greatly from state to state and sometimes even from county to county.

New regulations are published with a preamble that provides background material, an explanation of the regulatory requirements, and the effective date(s) of the regulations. This information is essential for understanding what you must do, why, and when. Federal regulations are first published in the *Federal Register* as preamble together with the regulations. Subsequently, the regulations (alone, without the preamble) are incorporated into the *Code of Federal Regulations*. For complete information, be sure to read the *Federal Register* publication of the final rule (i.e., the final regulation).

When regulations go into effect, they establish various requirements, and compliance with these regulatory requirements is mandatory. It is therefore essential (1) to know whether you are subject to any regulations that affect infectious and medical waste management practices, (2) to understand thoroughly the requirements of any such regulations that pertain to you, and (3) to comply with all the regulatory requirements.

Federal Regulation of Infectious Waste

Two statutes passed by Congress grant EPA the authority to regulate infectious waste disposal. These are the Resource Conservation and Recovery Act of 1976 (RCRA) and the Medical Waste Tracking Act of 1988 (MWTA).

Federal Regulation of Infectious Waste Under RCRA

Under RCRA, EPA can regulate the management and disposal of infectious wastes as hazardous waste. Although EPA did propose listing infectious wastes as hazardous wastes in 1978,* EPA has chosen *not* to issue final regulations because it does not perceive the problems of infectious waste disposal to be severe enough to warrant their regulation as hazardous waste at this time.

Instead of regulating infectious waste disposal, EPA issued a guidance manual on infectious waste management in 1982.[1] The guidance manual was later revised.[2] It must be stressed that these are *guidelines*, not regulations; therefore, they are recommendations, not legal requirements.

Federal Regulation of Medical Waste Under MWTA

Congress passed the MWTA in response to washups of medical wastes onto beaches along the Atlantic coast and the Great Lakes during 1988. The

---

*For EPA proposed rules for infectious waste management, see *Federal Register,* 43(243):58946–59028, December 18, 1978, especially pp. 58957–58958 and pp. 58963–58964. Although EPA never issued final regulations under RCRA for listing and managing infectious wastes as hazardous wastes, these rules remain proposed.

MWTA required EPA to establish a two-year demonstration tracking program to determine the effectiveness of tracking regulations in reducing the threat posed by medical wastes to human health and the environment. EPA issued medical waste tracking regulations in March 1989.[3] See Appendix A. (Types of regulated medical wastes are also discussed in Chapter 3, other regulatory requirements in Chapter 4.)

The medical waste tracking regulations apply to specified classes of regulated medical wastes that are generated in certain states. They impose certain packaging requirements as well as requiring the tracking of medical wastes from point of generation to point of disposal (or treatment and destruction). These federal regulations (or the equivalent state regulations) are in effect in Connecticut, New Jersey, New York, Rhode Island, and Puerto Rico.

The demonstration program (and regulations) will be in effect from June 22, 1989 to June 22, 1991. The data gathered during this program will be used in the decision on whether to make permanent the medical waste tracking regulations and whether to extend them to all the states.

Federal Regulation of Other Medical Wastes

Other medical wastes besides infectious waste may be subject to federal and/or state regulations. These include hazardous chemical wastes, radioactive wastes, and wastes with multiple hazards (i.e., any combination of biological, chemical, and radioactive hazards).

The disposal of many chemical wastes is regulated by the EPA under RCRA. RCRA regulations* impose stringent requirements on the management of hazardous chemical wastes using a cradle-to-grave approach that regulates the storage, transport, treatment, and disposal of these wastes. There are also record-keeping and reporting requirements that are intended to provide a paper trail for hazardous wastes from generation through final disposal. Many medical wastes are hazardous chemical waste, and, therefore, their management and disposal are regulated under RCRA. (See Chapter 11 for details.)

Disposal of all radioactive wastes is regulated by the Nuclear Regulatory Commission (NRC). The NRC regulations† impose requirements for the disposal of radioactive wastes. Many medical wastes have low levels of radioactivity, and they can be disposed of more easily than other wastes with higher levels of radioactivity. (See Chapter 12 for details.)

The management of wastes with multiple hazards has been more difficult because of jurisdictional disputes among the regulatory agencies. The EPA regulates infectious and hazardous wastes while the NRC regulates radioactive wastes. Because of conflicting regulatory requirements, generators of multihazardous

---

*For RCRA regulations, see *Code of Federal Regulations,* Title 40: Parts 260–280.

†For the NRC regulations that pertain to low-level radioactive wastes, see *Code of Federal Regulations,* Title 10: Parts 19 and 20. See especially Sections 20.301–20.311.

waste are often confronted with a dilemma in disposing of these wastes. (See Chapter 13 for more information.)

Other Federal Regulations Relevant to Waste Management

The Occupational Safety and Health Administration (OSHA) has always used as an enforcement tool the General Duty Clause* which requires an employer to "furnish . . . employment and a place of employment which are free from recognized hazards that are causing or likely to cause death or serious physical harm to his employees." Therefore, in the absence of regulations pertaining to a specific hazard, the General Duty Clause still requires safe employment and a safe place of employment. OSHA has not yet issued any regulations designed specifically to protect handlers of infectious waste. In 1987, OSHA did respond to the expressed concern of health care workers about their risk of occupational exposure to the human immunodeficiency virus by publishing recommendations in a Joint Advisory Notice [4,5] issued by the Departments of Labor and of Health and Human Services. It addresses the subject of protection against occupational exposure to the hepatitis B and human immunodeficiency viruses. The guidelines are relevant and applicable to health-care workers as well as to handlers of infectious wastes. The recommendations include:

- classification of each job-related task according to the degree of risk for exposure to blood, body fluids, or tissues
- identification of those workers who are at risk for exposure on a routine or occasional basis
- development of standard operating procedures (SOPs) for each high-risk task
- provision of training and education
- provision of engineering controls and personal protective equipment to minimize exposure
- monitoring of the effectiveness of the program

OSHA is now converting these recommendations into regulatory requirements, and proposed regulations were published in May 1989.[6] The proposed rule also includes sections that would impose regulatory requirements for the packaging and handling of infectious wastes. It must be stressed that the proposed rule provides information on current OSHA thinking, but the final regulations may or may not be identical to the proposal. Therefore, it is advisable to consider the proposed rule and to follow the recommendations in the Joint Advisory Notice until the final regulatory requirements are published in the *Federal Register*.

---

*Section 5(a)(1) of the Occupational Safety and Health Act of 1970.

The OSHA hazard communication/right-to-know regulations[7] were originally written regarding employee exposure to hazardous chemicals in the manufacturing industries, but they now apply to all industries. These regulations require employers to inform their employees about hazards that are present in the workplace.

Although the hazard communication regulations pertain to hazardous chemicals, it is prudent to extend the right-to-know concept to all hazards in the workplace. Workers who handle infectious and medical wastes are at risk for the hazards posed by exposure to these wastes, and they should be informed of these hazards under the OSHA right-to-know program. (See Chapter 14 for details.)

State Regulations

The states have taken different approaches to the regulation of infectious wastes. Infectious wastes have been regulated as solid wastes, special wastes, or hazardous wastes. Most states have at least some regulations pertaining to the disposal of infectious wastes. These regulations usually prohibit the landfilling of untreated infectious waste and specify which treatment/disposal techniques must be used. Other common regulatory provisions are those relating to storage conditions and transportation of the waste. In addition, some states have infectious waste tracking regulations.

State regulations usually apply only to infectious wastes that are generated within the particular state. Some states, however, also regulate infectious wastes that are brought into the state for treatment and/or disposal. Many states are now developing new infectious waste regulations or revising their current regulations. Therefore, state regulation of infectious waste is in a perpetual state of flux.

State regulations for the management of other medical wastes (e.g., hazardous chemical wastes) must be at least as stringent as the comparable federal regulations. In some states they are even more stringent.

You must know which regulations are in effect in your state, and you must understand all the regulatory requirements that apply to and affect your management of the various infectious and medical wastes that are generated at your institution or facility.

Local Regulations

Local regulations (or "ordinances" and "codes," as they are also called) are issued by local governments such as counties and municipalities. Local ordinances often specify which types of waste may or may not be deposited in the local landfill or burned in the municipal/county incinerator. There are also restrictions on the constituents and other characteristics of wastewater that is discharged to the sanitary sewer system. In addition, some localities have regulated medical waste incineration.

You must know which local ordinances affect your infectious waste management practices, and you must understand and comply with these regulatory requirements.

STANDARDS

Standards are usually established by professional organizations. Only governmental agencies can write regulations, so standards are not regulations. Standards do not set regulatory requirements unless they have been incorporated into regulations. You must remember, though, that standards often serve other purposes, and you may have to meet established standards in order to achieve some type of certification.

The standards affecting infectious and medical waste management are usually incidental to some other, primary purpose of the organization that sets the standards. Two organizations with standards that are relevant to infectious and medical waste management are the Joint Commission on the Accreditation of Healthcare Organizations (JCAHO) and the American Society for Testing and Materials (ASTM).

JCAHO Standards

The Joint Commission on Accreditation of Healthcare Organizations is the organization that accredits American hospitals, and it establishes the standards that hospitals must meet in order to achieve JCAHO accreditation. These standards are published annually in the *Accreditation Manual for Hospitals.*[8]

Among the numerous JCAHO standards is one pertaining to the management of infectious waste and hazardous waste. This is JCAHO Standard #PL.1.10, which states that a hospital must have "a hazardous materials and wastes program, designed and operated in accordance with applicable law and regulation, to identify and control hazardous materials and wastes."[9] To quote from JCAHO Standard #PL.1.10, this program must include the following elements:

- Policies and procedures for identifying, handling, storing, using, and disposing of . . . hazardous wastes from generation to final disposal;
- Training for and, as appropriate, monitoring of personnel who manage and/or regularly come into contact with hazardous materials and/or wastes;
- Monitoring of compliance with the program's requirements; and
- Evaluation of the effectiveness of the program, with reports to the safety committee and to those responsible for monitoring activities.[9]

If you are managing the infectious and other medical wastes produced in a hospital, your waste management system should meet the requirements of JCAHO

Standard #PL.1.10. Failure to do so could mean deficiencies during the accreditation inspection and even the risk of loss of accreditation. Following the requirements of this standard will help to ensure that you establish and maintain good systems for managing your infectious and medical wastes.

ASTM Standards

ASTM is one of the principal standard-setting organizations in the United States, and the standards that it adopts usually become industry standards. Most ASTM standards relate to materials and procedures.

There currently are two ASTM standards relevant to infectious and medical waste management. These are standards for testing the strength of plastic through use of a dart test.[10] The dart test standard has been incorporated into some regulations; for example, California regulations establish this test as a criterion for the plastic material in plastic bags that are used to contain infectious waste.

Two recently formed ASTM subcommittees are now studying the need for and the possibility of developing additional standards that relate to the management of infectious wastes. One subcommittee is developing a standard for puncture resistance in sharps containers;* the other is considering standards for red bags and for various infectious and medical waste treatment technologies.†

Development of ASTM standards is a time-consuming process. Therefore, it will probably be some time before any of these standards are adopted and then become industry standards.

GUIDELINES

Various federal and state government agencies and professional organizations have issued guidelines that were written specifically for infectious waste management or include sections on this subject. Guidelines have the status of recommendations, and no guideline is mandatory unless it is specifically incorporated into a regulation.

Federal Guidelines for Infectious Waste Management

The most recent EPA guidelines were published in 1986 as "EPA Guide for Infectious Waste Management."[2] These guidelines include the EPA definition and classification of infectious waste as well as recommendations for the handling, packaging, storage, movement, treatment, and disposal of the different

---

*ASTM Subcommittee F-04.08.01 on Puncture Resistance in Needle Disposal Containers.
†ASTM Subcommittee D-34.17 on Medical Waste.

types of infectious waste. These recommendations are referenced in the medical waste tracking regulations[3] for use in those aspects of infectious waste management that are not addressed by the regulations.

The Centers for Disease Control (CDC) of the United States Public Health Service has included recommendations for management of "infective" waste in several of its guidelines. The most recent such publication is "Recommendations for Prevention of HIV Transmission in Health-Care Settings."[11] Management of infective waste is also addressed in the CDC "Guideline for Handwashing and Hospital Environmental Control."[12]

The National Institute for Occupational Safety and Health (NIOSH) within the U.S. Department of Health and Human Services has issued guidelines for protecting the safety and health of health care workers.[13] The recommendations include a chapter on the handling of medical wastes. OSHA published "Guidelines for Cytotoxic (Antineoplastic) Drugs."[14] The guidelines include recommendations for disposal of these medical wastes.

State Guidelines for Infectious Waste Management

A few states have issued guidelines for infectious waste management rather than regulations. In these states, the intent was to provide guidance on the interpretation and implementation of legislation that was enacted to regulate the management of infectious waste.

Other Guidelines

The National Committee for Clinical Laboratory Standards is developing guidelines on protecting laboratory workers from infectious disease transmitted by blood and tissue.[15] The section on waste management refers to the EPA guidelines.

COMMUNITY STANDARDS

It is important to recognize that a community — and the courts — may regard standards and guidelines as the accepted techniques for management of infectious wastes even though they do not have the legal standing of regulatory requirements. That is, standards and guidelines that are issued by government agencies and professional organizations may be perceived as having the status of community standards. If recommendations are generally accepted in the industry and in the community, then it is expected that generators and other handlers of infectious waste should and will follow these recommendations. Someone who does not do so may be perceived as negligent for not acting in accordance with generally accepted practices.

REFERENCES

1. U.S. Environmental Protection Agency. "Draft Manual for Infectious Waste Management." SW–957. U.S. EPA, Washington, DC (September 1982).
2. U.S. Environmental Protection Agency. "EPA Guide for Infectious Waste Management." EPA/530-SW-86-014. U.S. EPA, Washington, DC (May 1986).
3. U.S. Environmental Protection Agency. "Standards for the Tracking and Management of Medical Waste; Interim Final Rule and Request for Comments." *Federal Register* 54(56):12326–12395 (March 24, 1989).
4. U.S. Department of Labor and U.S. Department of Health and Human Services. "Joint Advisory Notice: Protection Against Occupational Exposure to Hepatitis B Virus (HBV) and Human Immunodeficiency Virus (HIV)." DOL/DHHS, Washington, DC (October 19, 1987).
5. U.S. Department of Labor and U.S. Department of Health and Human Services. "Joint Advisory Notice: Protection Against Occupational Exposure to Hepatitis B Virus (HBV) and Human Immunodeficiency Virus (HIV)." *Federal Register* 52(210):41818–41824 (October 30, 1987).
6. U.S. Department of Labor, Occupational Safety and Health Administration. "Occupational Exposure to Bloodborne Pathogens; Proposed Rule and Notice of Hearing." *Federal Register* 54(102):23042–23139 (May 30, 1989).
7. *Code of Federal Regulations* Title 29, Section 1910.1200, "Hazard Communication."
8. Joint Commission on Accreditation of Healthcare Organizations. *Accreditation Manual for Hospitals*, 1990 ed. (Chicago: JCAHO, 1989).
9. Joint Commission on Accreditation of Healthcare Organizations. Standard #PL.1.10: "Hazardous Materials and Wastes," in chapter on Plant, Technology, and Safety Management, in *Accreditation Manual for Hospitals*, 1990 ed. (Chicago: JCAHO, 1989).
10. American Society for Testing and Materials. ASTM Standard #D1709–85: "Standard Test Methods for Impact Resistance of Polyethylene Film by the Free-Falling Dart Method," and ASTM Standard #D4272–85: "Standard Test Method for Impact Resistance for Plastic Film by the Instrumented Dart Drop Method" (Philadelphia: ASTM).
11. Centers for Disease Control (U.S. Department of Health and Human Services, Public Health Service). "Recommendations for Prevention of HIV Transmission in Health-Care Settings." *Morbidity and Mortality Weekly Report* 36 (suppl. #2S):1S–18S (August 21, 1987).
12. Garner, J. S. and M. S. Favero. "CDC Guidelines for the Prevention and Control of Nosocomial Infections. Guideline for Handwashing and Hospital Environmental Control, 1985." DHHS Publication #99–1117. U.S. Department of Health and Human Services, Public Health Service, Centers for Disease Control, Atlanta, GA (1985).
13. National Institute for Occupational Safety and Health (U.S. Department of Health and Human Services). "Guidelines for Protecting the Safety and Health of Health-Care Workers." DHHS (NIOSH) Publication #88–119 (September 1988).
14. Occupational Safety and Health Administration. "Guidelines for Cytotoxic (Antineoplastic) Drugs." OSHA Instruction Publication 8-1.1. OSHA, Office of Occupational Medicine (January 29, 1986).
15. National Committee for Clinical Laboratory Standards. "Protection of Laboratory Workers from Infectious Disease Transmitted by Blood and Tissue: Tentative Guideline." Publication #M29T. NCCLS, Villanova, PA (1989).

SECTION II

Infectious Waste Management

CHAPTER 3

Identification of Infectious Waste

It is essential, first of all, to define what is meant by "infectious waste." At present, there is not even agreement on the terminology that should be used for this type of waste. Various terms are being used more or less synonymously with infectious waste, including biohazardous waste, biological waste, medical waste, hospital waste, medical hazardous waste, infective waste, microbiological waste, pathological waste, and red bag waste.

The term "infectious waste" is used throughout this book. It denotes waste that is capable of producing an infectious disease, in accordance with the EPA definition of the term.[1]

There are four possible routes of disease transmission, i.e., ways in which infectious agents can enter the body to cause infectious disease. These are:

- through the skin via broken skin, cuts, scrapes, or puncture wounds
- through mucous membranes via splashing onto the mucous membranes of the eyes, nose, or mouth
- by inhalation
- by ingestion

Exposure to infectious agents present in infectious waste could result in disease transmission by any of these routes, depending on the type of exposure.

IDENTIFYING INFECTIOUS WASTE

The identification of infectious waste is a troublesome issue because there is no consensus about which wastes should be managed as infectious. The difficulty in establishing what is infectious waste results in part from lack of data. Of the information in the literature about the microbial load of different types of waste, little has been derived from carefully designed and conducted scientific studies. There are also few published reports of illnesses that could be specifically attributed to the mismanagement of infectious waste. The data that are available pertain only to illness caused by needle stick injuries.

In consideration of the lack of proof that infectious waste is really a problem, what is the best approach to take? We do *not* recommend the culturing of wastes to determine which, if any, infectious agents are present—the results cannot be reliable because of the uncertainties connected with obtaining a representative sample and using proper culture conditions. The procedure also entails risk of exposure and considerable expense. Therefore, there is no justification for culturing waste in order to determine if it is infectious. Besides, there are no accepted criteria (in terms of species and numbers of microorganisms) that define when a waste is "infectious."

Guidance provided by professional organizations and federal agencies is vague and therefore not very helpful if you are seeking specific recommendations. Standards published by the JCAHO state only that there should be a program for managing infectious waste and that the hospital must determine which wastes are infectious.[2]

Guidelines published by different federal agencies—that is, the EPA[3] and the CDC[4-7]—recommend different categories of infectious waste. OSHA proposed regulations for occupational exposure to bloodborne pathogens define infectious wastes.[8] EPA regulations[9] developed for the medical waste tracking demonstration program include yet another list: classes of regulated medical waste. Other lists of infectious waste categories appear in the literature.[10]

The categories of infective waste designated by the CDC are listed in Table 3.1. The infectious waste categories listed by EPA in guidance and in regulations are in Table 3.2.

In addition to the problem of differing recommendations, each published guideline states or implies that the final decision about classifying waste as infectious should be made only by someone who has sufficient knowledge to do so. In other

Table 3.1. CDC Recommended Categories of Infective Waste

Category	CDC[4]	CDC[6,7]
Microbiological laboratory waste	yes	yes
Pathology waste	yes	yes
Blood specimens and blood products	yes	yes
Sharps	no	yes
Isolation waste	no	yes

Table 3.2. EPA Categories of Infectious Waste

Category	Recommended Category[1]	Class of Regulated Medical Waste[9]
Isolation wastes	yes	yes
Cultures and stocks and associated biologicals	yes	yes[a]
Human blood and blood products	yes	yes[b]
Pathological wastes	yes	yes[c]
Contaminated sharps	yes	yes[d]
Contaminated animal carcasses, body parts, and bedding	yes	yes
Wastes from surgery and autopsy	optional	maybe[e]
Contaminated laboratory wastes	optional	maybe[f]
Dialysis unit wastes	optional	maybe[g]
Contaminated equipment	optional	no
Unused sharps	no	yes

[a]Includes cultures and devices used to transfer, inoculate, and mix cultures.
[b]Includes items that are or were saturated and/or dripping with blood, blood containers, and intravenous bags.
[c]Includes specimens of body fluids and their containers.
[d]Includes culture dishes with or without infectious agents, and slides and cover slips that were in contact with infectious agents.
[e]Wastes in this category may be included in other classes of regulated medical waste. (See footnotes b and c to this table)
[f]Wastes in this category may be included in other classes of regulated medical waste. (See footnotes a–d to this table)
[g]Wastes in this category may be included in another class of regulated medical waste. (See footnote b to this table)

words, for many types of waste the decision must be based on assessment of the particular risks involved rather than on specific guidance or scientific evidence. Infection control practitioners can help make this judgment.

The best approach is to address the issue from the aspect of the *potential* infectiousness of the waste. Evaluate each type of waste to determine whether it should be managed as infectious. In the decisionmaking process, ask the following questions:

- What type of waste is it?
- What is the likelihood that the waste contains infectious agents? Is the waste potentially infectious?
- What risk is potentially inherent in the waste? What risk do these infectious agents pose to health care workers? To sanitation workers? To operators

of waste treatment equipment? To waste haulers? To landfill workers? To the public?

- What is the likelihood that someone will be exposed to the waste if it is managed as general trash? If it is managed as infectious waste?
- Will management of the waste as infectious waste eliminate or reduce this risk?

It is best to make the decisions on the basis of the potential risk inherent in the waste. If the waste could be infectious, it is prudent to manage it as infectious.

TYPES OF INFECTIOUS WASTE

Our recommendations for types of infectious waste evolved from consideration of the various guidelines of CDC and EPA for infectious waste categories and optional categories, as well as EPA's list of regulated medical wastes. Our list includes the CDC and EPA categories of infectious waste. It is, we think, a more logical grouping of waste types that is suitable for practical application in the hospital, laboratory, and other places where infectious wastes are generated.

We recommend that the following types of waste be classified and managed as infectious waste:

- human blood and blood products
- cultures and stocks of infectious agents
- pathological wastes
- contaminated sharps*
- contaminated laboratory wastes
- contaminated wastes from patient care
- discarded biologicals
- contaminated animal carcasses, body parts, and bedding
- contaminated equipment
- miscellaneous infectious wastes

The rest of this chapter is a discussion of these infectious waste categories. For each category, there is a discussion of the types of waste included and appropriate management and treatment techniques.

Human Blood and Blood Products

Waste human blood and blood products should always be classified and managed as infectious waste because of the possible presence of infectious agents that cause

---

*From the management perspective, it is best to manage all sharps uniformly, without differentiating between contaminated and noncontaminated items.

blood-borne disease. A major concern today is acquired immune deficiency syndrome (AIDS), although there are also other blood-borne diseases—especially hepatitis B—that are serious, debilitating, and sometimes fatal. Because of the risk that blood-borne diseases could be transmitted by exposure to blood, it is essential that measures be taken to minimize such exposures.[8,11] This includes minimizing the risk of exposure to blood and blood products in the waste.

Wastes in this category include bulk blood and blood products as well as smaller quantities such as the blood samples that are drawn for testing. Since these wastes are liquid, small quantities of blood are best managed in terms of the associated container or fomites. Blood-contaminated test tubes, capillary tubes, and microscope slides and cover slips should be handled together with other sharps. Blood-soaked bandages fit logically into the category of contaminated wastes from patient care.

Waste human blood is best treated by steam sterilization or incineration. After sterilization, the liquid portion can be safely poured off into a drain. For management of untreated blood, evaluate the relative hazards of disposal into the sewer system and of movement through the facility. The first option has the risk of exposure due to splashing and aerosolization while the latter option has the risk of spills with potential for exposure during the spill and its cleanup.

Cultures and Stocks of Infectious Agents

Cultures and stocks of infectious agents should always be managed as infectious wastes because they contain large numbers of infectious microorganisms at high concentrations. The risk is inherent in cultures from medical, research, and industrial laboratories. All waste cultures and stocks of infectious agents should be managed as infectious waste.

It has been general practice to sterilize cultures from medical microbiology laboratories before disposal. This is done because of the types of infectious agents that are cultured and the high concentrations of microorganisms present in these cultures. Steam sterilizers in the laboratory are usually used for sterilizing these cultures.

One question that may arise is whether these wastes should be sterilized in the laboratory, elsewhere within the facility, or at an offsite infectious waste treatment plant. It is best to sterilize cultures onsite and right in the laboratory, which has been common practice. This approach eliminates risk of exposure during movement of untreated wastes through the facility as well as en route to an offsite treatment plant. This topic and recommended precautions for use when untreated wastes must be moved are addressed in the guidelines issued jointly by the Centers for Disease Control and the National Institutes of Health (NIH), both within the Public Health Service of the U.S. Department of Health and Human Services.[12]

Selection of the appropriate management alternative for your particular facility must be based on evaluation of the various aspects of the situation. The following questions are relevant:

- Which infectious agents are present in the waste cultures and stocks? Are they virulent? Are the diseases that they cause severe? Are they preventable? Are they treatable?
- Are the infectious agents present in the cultures at high concentrations?
- What type of treatment is most appropriate for this type of waste?
- What equipment is available for sterilizing the wastes? Is it located in the laboratory? Does it have sufficient capacity for treating all the waste cultures that are generated in the laboratory?
- If there is not sufficient treatment capacity in the laboratory, is other equipment available elsewhere in the facility? Can the wastes be moved safely and easily to this location for treatment?
- If the facility has no treatment capability, is there an offsite infectious waste treatment and disposal option? Does the alternative provide adequate procedures and safeguards to ensure the safe handling of this type of waste?

Pathological Wastes

Pathological wastes are body tissues that are removed during surgery or autopsy. This category includes tissue samples removed during biopsy, body tissues and organs, amputated limbs, and body fluids. Special handling of pathological wastes is warranted for two reasons: the infectious potential of the body tissues and aesthetic considerations.

Religious beliefs sometimes become the determining factor in the decision on how to handle these wastes. In some religions, it is important to bury the entire body, including any body parts that may have been amputated or otherwise removed surgically. For patients with such beliefs, the body parts should be made available to the patient or the patient's family for burial by a mortician.

Otherwise, it is the responsibility of the hospital to dispose of body parts. Incineration is often the method of choice for pathological wastes because this technique takes care simultaneously of both potential infectiousness and aesthetics. Steam sterilization, however, leaves the pathological wastes intact, and the problem of aesthetics remains. It is not acceptable that recognizable body parts be placed in a landfill. Therefore, when pathological wastes are steam sterilized, additional processing is necessary before disposal; options include incineration and grinding of the sterilized wastes.

Another acceptable management alternative for pathological wastes is handling by a mortician who provides cremation or burial. Many hospitals, especially those without access to an incinerator, routinely use the services of a mortician for the disposal of pathological wastes.

Contaminated Sharps

The category of sharps includes hypodermic needles and syringes, intravenous needles, scalpel blades, lances, disposable pipettes, capillary tubes, microscope slides and cover slips, and broken glass. Contaminated sharps are now universally recognized as a type of waste that requires special handling because of the double hazard involved—risk of injury and risk of disease. Cuts, scrapes, and puncture wounds from contaminated sharps can transmit infectious agents through the skin. The injured person is then at risk for infection and disease, including blood-borne diseases such as AIDS and hepatitis B.

Sharps are a known occupational hazard to all handlers of medical waste. Sharps that are tossed loosely in with the trash endanger all who subsequently handle that waste. There is only one way to prevent injury from sharps; that is to discard them directly into special containers that confine the sharps and protect against injury. Use of special containers for discarded sharps would be required by the proposed OSHA rule.[8]

Since publication of the original EPA infectious waste management guidelines[3] in 1982, many products were marketed as sharps containers, that is, special containers for sharp wastes. Common features of sharps containers are rigidity, puncture resistance to retard needle penetration through the wall, narrow openings to prevent retrieval of discarded needles and syringes, and a lockable lid or some other type of locking mechanism. Specific characteristics of the various sharps containers differ, and you should evaluate available products to ascertain which one best meets your needs. An ASTM standard for puncture resistance in sharps containers is now under development;* once adopted, this standard should bring some uniformity to the puncture-resistant quality of sharps containers.

In addition to occupational safety and health concerns regarding contaminated sharps, there are also public health considerations. One concern is the risk to the public from needles and syringes that appear in public places and degrade the environment.† Another concern relates to drug abuse and the availability of needles and syringes. Proper management of sharps can alleviate both concerns. Use of suitable containers eliminates loose needles and minimizes the risk they pose of needle sticks and other injuries. Appropriate treatment of sharps makes the needles and syringes unsuitable for reuse.

Good management policy for sharps eliminates the two hazards in sharps (infectiousness and physical injury) and prevents their reuse. Incineration accomplishes these goals simultaneously. After steam sterilization, an additional step (such as grinding or incineration) is necessary because the needles are still usable even though they are no longer infectious.

Sharps that were not exposed to infectious agents are not contaminated and, therefore, are not infectious per se. Nevertheless, we strongly recommend treating *all* sharps *uniformly* without distinguishing between those that are contaminated and those that are not.‡

---

*ASTM Subcommittee #F-04.08.01 on Puncture Resistance in Needle Disposal Containers.

†For example, medical wastes that washed onto beaches along the Atlantic Ocean, the Great Lakes, and the Gulf of Mexico during the summer of 1988 included needles and syringes.

‡EPA's medical waste tracking regulations establish two categories of regulated medical wastes for sharps: Sharps and Unused Sharps. See Reference #9.

There are several reasons for adopting a uniform policy for all sharps:

- Although uncontaminated sharps are much less likely to cause disease than contaminated sharps, there remains the risk of physical injury (cuts, scrapes, and needle sticks).
- Risk of infection accompanies physical injury by sharps. Even a sterile sharp discarded into waste becomes nonsterile from being in the waste.
- No one likes to be stuck, and physical injury from sharps is unpleasant. It is also disturbing to the injured person because of the fear of AIDS it often evokes. The person is hardly reassured by being told that the particular sharp that caused the injury was not contaminated because it had not been used on an infected person.
- A uniform sharps policy eliminates decisionmaking because no one has to decide whether or not a particular sharp is contaminated.
- Training and management are simpler, easier, and more efficient when all sharps are handled in exactly the same way.
- Uniform sharps handling means universal use of sharps containers, a practice that offers protection for all handlers of sharps.
- A uniform sharps handling policy is consistent with public health concerns about drug abuse and the reuse of needles and syringes.

Contaminated Laboratory Wastes

The category of contaminated laboratory wastes includes all potentially infectious wastes generated in microbiological, pathological, medical, research, or industrial laboratories that are not classified as another type of infectious waste (such as cultures, blood, or sharps). Before the advent of disposable laboratory supplies, used materials were routinely sterilized before reuse. With the introduction of disposables, these items became wastes that were generally discarded directly into the trash. In 1978, an article by Kaye[13] alerted users to the problem that many wastes from medical laboratories were contaminated.

Contaminated laboratory waste is one of the waste categories termed optionally infectious by EPA.[1] All laboratory wastes should be evaluated for potential infectiousness using professional judgment. Those wastes that are determined to be potentially infectious should be managed as infectious waste.

Examples of contaminated laboratory wastes include:

- specimen and culture containers such as cups, bottles, flasks, petri dishes, and test tubes
- implements used to manipulate infectious materials such as specimens and cultures (for example, swabs, spreaders, and pipettes)
- components of diagnostic kits that are contaminated by use with specimens

- personal protective equipment such as disposable gloves, lab coats, aprons, and masks that are grossly contaminated with blood, body fluids, secretions, excretions, or cultures

Some of these items belong in other infectious waste categories. For example, sharps from laboratories should be handled in accordance with the special procedures established for handling all sharps, and cultures should be managed like wastes in that particular category.

Contaminated laboratory wastes can be treated by any treatment technology that provides effective treatment of the waste. (See Chapters 5 through 8.) No particular treatment technique is best for all of these wastes.

Contaminated Wastes from Patient Care

Contaminated wastes from patient care originate from various areas of patient care other than laboratories, such as clinics, patient rooms, emergency rooms, operating rooms, hemodialysis centers, and morgues. This category includes all potentially infectious wastes that do not belong to other specific categories such as blood and body fluids, cultures, or sharps.

A general guide for determining which wastes from patient care are potentially infectious is to include in this category all wastes that are grossly contaminated with blood, body fluids, excretions, and secretions. This policy is consistent with the CDC recommendations for universal precautions.[4,5]

Wastes from general patient care include such diverse items as diapers and bed pads, intravenous tubing, catheters and bags, drainage tubing and pouches, and wet dressings (especially those soiled with blood, pus, or body fluids).[*]

Current management practices for wastes from surgery and autopsy vary greatly. In some hospitals, all wastes from surgery are handled as infectious. In others, surgical cases are classified as either "clean" or "dirty" and only the wastes from dirty cases are considered infectious. In other hospitals, only selected wastes from all surgery are managed as infectious. Obviously, the decision is an individual one that is affected by the outlook and perceptions of risk at each individual hospital.

Wastes from hemodialysis that have been grossly soiled by contact with patient blood should be managed as infectious waste. This includes disposable tubing, filters, towels, aprons, gloves, and lab coats.

Classifying wastes from patient care as infectious or not infectious is a decision based on judgment. There are three alternatives:

---

[*]Wastes such as blood and body fluids and hypodermic and intravenous needles and lances are also infectious wastes generated during patient care. They should, however, be classified as "blood and body fluids" and "sharps" respectively and receive the special handling designed for these types of infectious waste.

- wastes that should never be managed as infectious
- wastes that should always be managed as infectious
- wastes that require individual evaluation

Table 3.3 provides a general guide for classifying wastes from patient care. The difference between "always infectious" and "sometimes infectious" depends on how you distinguish between "grossly contaminated" and "slightly contaminated." It is a matter of personal evaluation and decisionmaking that requires professional judgment; input from the infection control practitioner is helpful. For more detailed guidance, see the discussion in the section entitled "Identifying Infectious Waste," earlier in this chapter.

Table 3.3. Guide to Classification of Wastes as Infectious or Not Infectious

Never infectious (regardless of the patient disease, the type of care, or the type of surgery)

- Outer wrappers of supplies
- Materials that were not in contact with patient blood, body fluids, secretions, or excretions

Always infectious

- Materials that are grossly soiled with patient bood, body fluids, secretions, or excretions

Sometimes infectious

- Materials that are lightly soiled with patient blood, body fluids, secretions, or excretions

Infectious wastes from patient care can be treated by any treatment technology that is effective. (See Chapters 5 through 8.) No particular treatment technique is best for all these wastes.

Discarded Biologicals

This category of infectious waste is comprised of waste biologicals, such as live and attenuated vaccines. Included are production wastes as well as products that are discarded for various reasons (quality control, recalls, outdating). Note that many wastes from production of biologicals are covered by other categories of infectious waste such as cultures and stocks of infectious agents, sharps, contaminated laboratory wastes, and contaminated animal carcasses, body parts, and bedding.

Discarded biologicals are usually incinerated. Incineration not only treats the infectiousness, it also destroys the material as well as the labels. These are important considerations for the manufacturer. (Incineration is discussed in Chapter 7.)

Production wastes, like other infectious wastes, can be treated by any technique that is effective. Heat sterilization and chemical disinfection are often appropriate for liquid wastes. (See Chapter 8.)

Contaminated Animal Carcasses, Body Parts, and Bedding

In some research projects, animals are infected with human pathogens in order to study disease processes and the efficacy and side effects of pharmaceuticals. The wastes from such research (that is, the animal carcasses, body parts, and bedding) are best handled as infectious waste so as to minimize the risk of exposure for waste handlers.

Animal carcasses and body parts should be handled like pathological waste (see above). Animal bedding can be difficult to treat—bedding material is a good insulator that can impede steam sterilization, and its high moisture content can prevent complete combustion during incineration. It is important to standardize and to validate treatment procedures and then to follow the established procedures in order to be certain that treatment will be effective.

Contaminated Equipment

Equipment and equipment parts are discarded when no longer useful because they cannot be repaired or they have become obsolete. Equipment may have been contaminated with infectious agents (for example, by spills and splashing) during use, and prudent management provides for decontamination of the equipment before it is actually thrown out. Therefore, from the safety perspective, contaminated equipment should be managed as infectious waste.

Usefulness is not a consideration with equipment that is being junked. If practical, steam sterilization or incineration could be used for treatment because the effect of steam or combustion does not matter when the equipment will not be used again. For contaminated equipment that is large (such as a centrifuge), the best approach may be to treat the object using formaldehyde decontamination before it is moved. (See Chapter 8.)

Miscellaneous Infectious Wastes

This category includes general types of infectious waste that are not readily assigned to another specific category. These wastes are usually generated in the handling of infectious materials and wastes.

Personal protective equipment should be worn whenever and wherever potentially infectious materials and wastes are handled. (Such use would be required by the proposed OSHA rule.[8]) These items are usually disposable. They include latex gloves, masks, aprons, and lab coats—some or all of which may be appropriate for use under particular circumstances.

Also in this category is the waste that is generated during cleanup of spills of infectious materials and wastes. These wastes include absorption materials (loose material as well as spill pillows), towels, mops, torn or broken containers, and the personal protective equipment that was worn during the cleanup.

All of these wastes should be managed as infectious. Any treatment technique that is effective may be used for treating these wastes. (See Chapters 5 through 8.)

REFERENCES

1. U.S. Environmental Protection Agency. "EPA Guide for Infectious Waste Management." EPA/530-SW-86-014. Washington, DC (May 1986).
2. Joint Commission on Accreditation of Healthcare Organizations. Standard #PL.1.10. "Hazardous Materials and Wastes," in chapter on Plant Technology, and Safety Management, in *Accreditation Manual for Hospitals,* 1990 ed. (Chicago: JCAHO, 1989).
3. U.S. Environmental Protection Agency. "Draft Manual for Infectious Waste Management." SW-957. U.S. EPA, Washington, DC (September 1982).
4. Centers for Disease Control. "Recommendations for Prevention of HIV Transmission in Health-Care Settings." *Morbidity and Mortality Weekly Report* 36 (suppl. #2S):1S-18S (August 21, 1987).
5. Centers for Disease Control. "Update: Universal Precautions for Prevention of Transmission of Human Immunodeficiency Virus, Hepatitis B Virus, and Other Bloodborne Pathogens in Health-Care Settings." *Morbidity and Mortality Weekly Report* 37(24):377-387 (June 24, 1988).
6. Garner, J. S. and M. S. Favero. "CDC Guidelines for the Prevention and Control of Nosocomial Infections. Guideline for Handwashing and Hospital Environmental Control." DHHS Publication #99-1117. U.S. Department of Health and Human Services, Public Health Service, Centers for Disease Control, Atlanta, GA (1985).
7. Garner, J. S. and B. P. Simmons. "CDC Guidelines for the Prevention and Control of Nosocomial Infections. Guideline for Isolation Precautions in Hospitals." DHHS Publication #(CDC)83-8314. U.S. Department of Health and Human Services, Public Health Service, Centers for Disease Control, Atlanta, GA (1983).
8. U.S. Department of Labor, Occupational Safety and Health Administration. "Occupational Exposure to Bloodborne Pathogens; Proposed Rule and Notice of Hearing." *Federal Register* 54(102):23042-23139 (May 30, 1989).
9. U.S. Environmental Protection Agency. "Standards for the Tracking and Management of Medical Waste; Interim Final Rule and Request for Comments." *Federal Register* 54(56):12326-12395 (March 24, 1989).
10. Liberman, D. F. and J. G. Gordon, Eds. *Biohazards Management Handbook* (New York: Marcel Dekker, Inc., 1989), ch. 12.
11. U.S. Department of Labor and U.S. Department of Health and Human Services. "Joint Advisory Notice: Protection Against Occupational Exposure to Hepatitis B Virus (HBV) and Human Immunodeficiency Virus (HIV)." DOL/DHHS, Washington, DC (October 19, 1987).
12. Centers for Disease Control and National Institutes of Health (U.S. Department of Health and Human Services, Public Health Service). "Biosafety in Microbiological and Biomedical Laboratories," 2nd ed. DHHS Publication #(NIH) 88-8395. Washington, DC (May 1988).
13. Kaye, S. "Disposing of Laboratory Disposables." *Laboratory Management* 16(6):37-44 (1978).

CHAPTER 4

Handling, Storage, and Transport
of Infectious Waste

The principal cause for concern during the handling of infectious wastes is the risk of exposure to infectious agents (pathogens) that may be present in this type of waste. This risk persists during all phases of infectious waste handling, that is, during waste discard, collection and movement within the facility, storage, onsite treatment, and transport offsite for treatment and disposal. Waste handlers have the greatest risk of exposure, but other persons may also be at risk.

The greatest hazard is from direct contact with sharps (items such as hypodermic and IV needles, razors and other blades, scalpels, and broken glass). Sharps can cause puncture wounds, scratches, and scrapes. When the skin is not intact—either because of an existing injury or other medical condition or as the result of a sharps-induced injury—infectious agents can penetrate the skin.

During waste handling, there is also potential for exposure through inhalation of pathogen-containing aerosols or dusts. Pathogens can also be ingested when a person eats or smokes with hands dirtied during handling of infectious wastes. Exposure can also occur through splashes of infectious liquids onto mucous membranes (such as those in the mouth, nose, and eyes).

When one understands the possible routes of exposure, it is apparent that the best way to minimize the risk of exposure is to ensure that the infectious waste is properly contained at all times. Some basic principles and procedures can help to achieve the goal of minimizing exposures. These are:

- packaging the waste properly
- maintaining the packaging and containment of the waste and avoiding practices that may tear or break waste containers
- avoiding physical contact with the waste

43

- using personal protective equipment (gloves, aprons, masks, goggles, etc.) as needed for particular tasks
- handling the waste as little as possible
- limiting the number of persons with potential for exposure
- avoiding spills and accidents

This chapter presents a detailed discussion of the various aspects of waste handling, from discard of the waste, through waste collection, storage, and treatment, to transport of infectious waste to an offsite treatment facility. Each section is devoted to a particular aspect of waste handling and discusses the different problems that may arise, the inherent risks, and the practices and precautions that will minimize these risks.

CONTAINERS AND PACKAGING

The most important element in minimizing the risk of handling infectious waste is the use of proper containers for holding the waste. It is best to discard infectious waste directly into the designated containers at the point of discard (that is, at the place where the material is discarded and thus becomes a waste). When this is done, subsequent direct handling of the waste such as sorting and repackaging—which is accompanied by risk of exposure—is not necessary and can be avoided.

Various factors must be considered in the selection of containers for infectious wastes. These factors include:

- the type of waste
- waste collection procedures
- waste handling practices
- waste storage
- waste treatment
- transport of waste offsite

The first consideration in selection of waste containers is the type of infectious waste. From the perspective of containment, infectious wastes are of three general types: sharps, solid wastes, and liquids. These types of waste differ greatly in their physical nature; consequently, there are different requirements for the kinds of containers and packaging that will adequately and safely contain each type of waste. In practice, therefore, at least three different types of containers should be used for infectious wastes.

Containment of Sharps

The use of special containers for sharps is necessary in order to protect waste handlers from injury caused by contact with the discarded sharps. Another factor to be considered is public health and drug abuse—no one should have easy access to discarded needles and syringes. In consideration of these two factors, sharps containers should have the following features:

- puncture resistance
- impermeability
- rigidity
- tamper resistance
- proper marking

The OSHA proposed rule[1] would require most of these features in sharps containers.

Puncture Resistance

Although there is not yet a universal definition or standard for "puncture resistance" in sharps containers,* the commonly accepted meaning of the term is the ability of the container to withstand punctures through the wall by the contained sharps during ordinary usage and handling. Puncture resistance is an essential feature for sharps containers in order to ensure that discarded sharps can be safely handled with minimum risk of exposure and injury.

When additional waste is pushed into a full sharps container, another sharp can be forced through the wall, thereby negating the puncture-resistant quality of the container. Therefore, personnel should never overfill sharps containers. This issue is also addressed in the OSHA proposed rule.[1]

Impermeability

Sharps containers should be impermeable during ordinary usage and handling so that residual liquid in syringes and other items will not leak from the containers. This feature is important to minimize the risk of exposure to infectious liquids leaking or spilling from sharps containers.

Rigidity

Rigidity is another important feature in sharps containers. This characteristic ensures that sharps containers retain their shape and puncture resistance during use and handling. Rigid containers can be easily stacked before use. They are handled easily on waste collection carts. Rigid sharps containers are easy to package in cartons if this procedure is used (either voluntarily or in response to the medical waste tracking regulations)[2] for transport of sharps away from the facility.

---

*An ASTM subcommittee, #F-04.08.01, is now developing an ASTM standard for materials to be used in the manufacture of puncture-resistant containers.

Tamper Resistance

Tamper resistance is essential in sharps containers to prevent drug abusers from having easy access to discarded needles and syringes. Many manufacturers of sharps containers have incorporated this feature into their products. Some sharps containers have small openings and/or locks that deter removal of material from containers that are in use. Some containers have lids that lock in place; these lids should be tightened and thereby locked when the container is full and ready to be collected.

Proper Marking

Proper marking is needed to clearly identify containers that hold infectious sharps. This is useful and important not only for the users (that is, those who discard sharps into the containers), but also for the waste handlers. Three types of markings are in common use: the red (or red-orange or orange) color that is commonly used to denote biohazards, the universal biohazard symbol, and wording such as "contaminated sharps."

Some states and localities have regulations that specify marking requirements for sharps containers. The particulars of these requirements are not uniform throughout the United States.

Failure to use sharps containers that are properly marked to designate the biohazard is now sufficient basis for a citation by OSHA inspectors. Acceptable markings include color, wording, and/or the biohazard symbol.

Selecting Sharps Containers

Many institutions are now opting for extensive use of sharps containers so that they are in place as close to the site of use as possible. The objective is to allow the user to discard the sharp directly into a sharps container without having to walk any distance. With this approach, sharps containers are situated in every patient room, on every phlebotomy cart, on every laboratory bench. Consider this approach and its value for your unique situation. It would be required by the OSHA proposed rule.[1]

Many different sharps containers are on the market. Most have the desired characteristics to a greater or lesser degree. Compare the various products. Select the product that best meets your particular needs.

Containment of Solid Wastes

Solid wastes are usually discarded into plastic bags. Certain features are essential in plastic bags to ensure that the waste will be contained without spilling and that there will be a clear indication of the type of waste that is within the bag. These features include the following:

- impermeability
- strength
- proper markings

These characteristics should be sought when plastic bags are purchased for use as containers of infectious waste. These features are also relevant in establishing standard operating procedures (SOPs) for how the plastic bags are used.

Impermeability

Solid infectious waste often contains some wet material (although plastic bags should not be used for liquids; see below). A person who handles a leaking or wet bag is especially at risk of exposure to infectious agents. In order to prevent spilling and leaking from plastic bags, the bags must be impermeable. A plastic bag is most likely to leak through the seam; therefore, the type of seam can determine the suitability of a particular plastic bag for use in containing infectious wastes. Be sure to compare the different products that are available to find the one that meets your needs.

It is important that wet and leaking bags be repackaged immediately in order to minimize the risk of exposure. The simplest and easiest procedure is to double bag—that is, to place the wet bag within another plastic bag, then to seal or otherwise securely fasten the outer bag.

Strength

Strength of the plastic is another important feature. Bags should be strong enough to hold the intended loads without breaking or bursting. Unfortunately, it is difficult to judge the strength of plastic bags because there is no uniformly accepted measure of the strength of plastic film.

One index of the strength of plastic is the dart test.[3] This type of test is cited in the regulatory requirements of certain states. In California, for example, the plastic material that is used to manufacture bags for infectious wastes must pass the falling dart test.

In other states, the requirement is worded in terms of the thickness of the plastic. This requirement (e.g., 1.5 or 3.0 mils) is not really reliable as a measure of bag strength. Strength does not depend solely on thickness; it is also a function of other factors, including the type of plastic, its purity, and the construction of the bag. Although thicker is usually stronger, two different plastic bags of the same thickness do not necessarily have the same strength.

From practical considerations, plastic bags should not be overloaded with too much waste. The contents of the bag should be limited to the weight and the volume of waste that the bag can hold without bursting. Similarly, in order to avoid tearing of the bag, no sharp objects should be placed directly into plastic bags. Prompt cleanup of any spills that might occur will help to minimize the risk of exposure. (See Chapter 15.)

Proper Markings

Plastic bags that are used for infectious wastes are usually red or red-orange in color (although some state regulations may require the use of one particular color). As a result, the term "red bags" refers to plastic bags that contain infectious waste. If the plastic itself is not one of these distinctive colors, these colors are usually used for the markings on the bag (that is, for the biohazard symbol and/or the wording).

One problem with the use of red bags for infectious wastes is that steam sterilization does not usually affect the red color, and most red bags remain red after they have been steam sterilized. This could create difficulties in distinguishing between treated and untreated waste that is contained in red bags. For a discussion of this potential problem, see the section entitled "Waste Treatment" in this chapter.

Containment of Liquids

There are various ways in which liquid infectious wastes can be safely contained to minimize the risk of exposure. Selection of a suitable method of containment depends on whether the liquid wastes must be moved for treatment, how this is done, how far they must be moved, and the treatment process.

Liquid wastes that must be moved through a facility to treatment equipment are best contained in bottles or similar rigid containers with secure closures such as screw caps or corks. The use of another secondary container such as a pail or box to hold the bottles provides an extra degree of security and containment for spills, and reduces the risk of exposure.

When liquid wastes are to be steam sterilized, the container must not interfere with the effectiveness of treatment. Bottle caps or stoppers must be loosened for the sterilization process so it is important that this can be accomplished easily within the sterilizer. Another consideration is heat conductivity, an important factor in steam sterilization. (See Chapter 6.) Because metal enhances heat conduction, metal containers (as either the primary or the secondary container) are preferable to those made of glass. Plastic containers should not be used for liquid wastes that will be steam sterilized because plastic is an insulator rather than a conductor of heat and it will impede the heating of the liquid.

For liquid wastes that are to be incinerated, plastic containers are best because they are readily combustible. The use of metal and glass containers for liquid wastes that will be incinerated should be avoided because these materials can cause problems during incineration as well as the formation of slag within the incinerator.

Liquid infectious wastes do not necessarily have to be moved for treatment. Liquid waste is sometimes generated within a tank or other large container. In such cases, it is often practical to treat the waste directly in the tank or through a closed system that leads to the sanitary sewer. See Chapter 8 for a discussion of heat treatment of infectious liquid wastes.

WASTE HANDLING PRACTICES

In the interests of safety and risk minimization, infectious waste should be handled as little as possible. This means discard of waste directly into the designated container so that subsequent sorting, handling, and repackaging are not necessary.

Similarly, waste collection and movement procedures should be as simple as possible. The less the waste is handled, the smaller is the risk of mishaps, waste spillage, and exposures.

Waste Collection

Certain precautions are necessary when infectious wastes are collected that do not even have to be considered during the collection of general solid waste. The concerns about packaging and containerization that are so important in the selection of suitable containers are also relevant in waste collection. In order to minimize the risk of exposure, it is essential to maintain the integrity of the packaging throughout the waste collection process. In other words, the objective is to avoid those situations and incidents that can result in actual or potential exposure to infectious agents—things such as torn bags, broken boxes, wet containers, leaks, and spills.

Carts are generally used to move infectious wastes to the treatment or storage area. It is best to use different types of carts to move the different kinds of waste containers. Bins are suitable for bagged waste, whereas a cart with level shelves is usually better for rigid waste containers (for example, sharps containers and boxes).

Whatever type of cart is used, it should be easy to move, cleanable, and easily disinfected. In order to avoid the risk of exposures and contamination, it is best to dedicate carts to infectious waste transport—that is, they should be used only for this purpose and never for other activities such as transporting food or supplies. With a color coding or labeling system, carts for infectious waste are readily identifiable.

The route of waste collection and movement within the facility is another factor to be considered in the context of safe handling of infectious waste and minimization of the risk of exposure. In the ideal situation, separate corridors and elevators would be used to move the wastes from the points of generation to a central collection point for treatment or storage. This, however, is rarely possible.

Therefore, procedures for waste collection should be developed with the ultimate goal of minimizing the possibility of exposure to the waste during its movement. The following procedures can help to achieve this goal:

1. Use collection carts that:
   - are appropriate for each type of waste container (e.g., plastic bags or sharps containers) in order to minimize spills;
   - are dedicated to the collection of infectious wastes so that they will not be used for moving other items (thus avoiding possible contamination of food and supplies);
   - are easily moved in order to prevent injury, accidents, and spills;
   - are easily cleaned and disinfected in order to provide clean carts for waste movement.
2. Schedule periodic routine cleaning and disinfection of the collection carts.
3. Coordinate collection schedules with waste generation rates, storage capacity, and treatment capacity.
4. Develop collection routes and schedules that minimize encounters with the waste and the potential for exposure.

Waste collection procedures are an important element in the concept of minimizing exposures because improper procedures can result in exposures. The best packaging can be destroyed if a container is handled in a way that causes, for example, the tearing of bags or the spilling of waste. Such incidents can occur when inappropriate carts are used. Another cause is the use of mechanical rather than manual means to move waste (e.g., gravity chutes and pneumatic tubes).*

Compacting of waste can also result in exposures if the compacting process creates dusts and aerosols. Only compactors that operate under negative pressure (thereby retaining aerosols and dusts within the unit) should be used for compaction of infectious waste. Such compactors are now under development and commercially available.

Waste Storage

It is generally accepted that infectious wastes should be treated as soon as possible after they are generated, and that storage prior to treatment is acceptable only when the waste cannot be treated immediately. The reasons for such a policy are primarily aesthetics—waste will putrefy at room temperature, and it will become unpleasant to handle. In addition, many microorganisms multiply rapidly at room temperature, and the biohazard increases accordingly.

Some state regulations do specify requirements for storage of infectious waste in terms of time and temperature. These regulations set limitations on the amount of time infectious waste may be stored at room temperature, under refrigeration, and in a freezer. In other states, there are no such restrictions. Know which regulations apply to you.

---

*It is advisable, when evaluating automatic guided vehicle systems, to include consideration of the effect of the system on the integrity of the packaging.

Various factors in waste storage are important to minimize risk and ensure safety. These include the following policies and procedures:

- secure closing or sealing of all stored containers of infectious waste
- dedicated waste storage areas so that there is no mixing of wastes with supplies and food in the same refrigerator, freezer, or other storage area
- limited access so that only authorized persons can enter the area
- labeling and/or posting of the storage area so that it is readily apparent what type of material is stored within

Containers of infectious waste should be maintained in an intact state in the storage area. This may involve special precautions when wastes are moved into and out of storage. This extra handling must be done without compromising the integrity of the waste packaging. The use of "secondary containment" may be warranted—cartons, pails, and double bagging can help to protect the primary container.

WASTE TREATMENT

The same principles of minimizing opportunities for exposure apply to the treatment aspects of waste management. In general, these involve maintaining the integrity of the packaging as much as possible and using personal protective equipment as necessary.

Incineration

When infectious waste is incinerated, it is important to maintain the integrity of the packages of waste while they are being handled. Many incineration systems provide for mechanized handling of the waste, such as use of a conveyor belt that moves cartons of waste directly to the incinerator. Systems that involve the handling of loose waste are not acceptable because they provide too much potential for creation of dusts and aerosols and for employee exposure; such systems should be avoided.

Steam Sterilization

When infectious waste is steam sterilized, it may be necessary to open the packages or containers to allow the steam to penetrate throughout the waste. When this procedure is necessary, it must be done carefully to prevent the creation and spreading of dusts and aerosols. It is best done, if possible, within the sterilizer rather than in the open room in order to contain whatever dusts and aerosols may be generated.

One problem with steam sterilization is the red color of infectious waste containers. Waste handlers are being trained to correlate the color red with infectious waste. Red containers imply infectious waste. Within the generating facility such color coding is advantageous—it serves to designate those wastes that are to be managed as infectious, that must receive special handling. However, the same red color that is so important within the facility becomes a detriment when the waste has been steam sterilized and is being sent offsite to be landfilled for disposal. A hauler or a landfill worker will assume that the waste within a red sharps container or a red bag full of waste is infectious; such a worker cannot be easily convinced that the waste has been sterilized and is therefore no longer a hazard.

What can be done about this problem? There are now several products on the market or under development that allow relatively easy differentiation between treated and untreated plastic bags of infectious waste. Some plastic bags change color when the plastic is exposed to steam—the entire bag may change from red to brown or black, or some markings may change to black, or special markings may appear. A bag that completely changes color is best because the red color disappears, and it is then not necessary to look for special markings in order to ascertain that the bag was indeed steam sterilized.

An alternative is to use a plastic bag that crumples and partially disintegrates when exposed to steam (e.g., a bag manufactured from polyethylene). With this type of plastic, it is apparent and obvious when the bag and its contents have been treated with steam. Since a crumpled bag cannot hold (contain) anything, this type of plastic bag must be placed within another container (such as a kraft paper bag to keep the waste from spilling within the steam sterilizer. There is then an intact secondary container that is not red holding the treated waste.

Compatibility of Containers with Treatment Process

Another important feature that must be considered in selecting containers is the compatibility of the container with the treatment process. This is important from two different aspects. The concerns are (1) that the containers could impede the effectiveness of treatment, and (2) that they could produce deleterious side effects from the treatment process. These two concerns are related to the specific types of treatment.

In order for steam sterilization to be effective, the steam must be able to penetrate throughout the waste that is within the plastic bag. Autoclavable bags are supposed to be good for steam sterilization because they have two desirable features: they allow free penetration of steam and they remain intact during and after treatment. Unfortunately, the term "autoclavable" is not used uniformly. Sometimes it refers to the ability of steam to penetrate the bag while at other times it is used to denote the resistance of the bag to melting in the autoclave; the term does not always imply both characteristics.

There have been reports that some supposedly autoclavable bags can deter steam penetration, and, when this happens, the treatment is not effective.[4] Other bags allow steam penetration but crumple when exposed to steam. Therefore, it is important to check for both features when selecting plastic bags that will be used to contain infectious waste before, during, and after the steam sterilization process.

The concern about deleterious side effects relates to the incineration of infectious wastes contained in plastic bags. The principal concern is that hydrochloric acid can be generated from chlorinated plastics during incineration. Hydrochloric acid is undesirable for two reasons: it is corrosive to the incinerator and it is emitted with the stack gases.* The chlorine content of infectious waste can be minimized by avoiding the use of plastic bags and other disposable products that are manufactured from polyvinyl chloride (PVC) and other chlorine-containing plastics.

The solution to these various concerns is simple. Before plastic bags are ordered, it is important to know what type of treatment will be used. If the waste will be steam sterilized, the bags must be permeable to steam. They must also remain intact during and after steam treatment, or else they must be used in conjunction with a secondary container. If the waste will be incinerated, the bags must be manufactured from nonchlorinated plastics such as polyethylene and polypropylene.

TRANSPORT OF INFECTIOUS WASTE OFFSITE FOR TREATMENT

Special factors must be considered when infectious waste is transported offsite from the generating facility for treatment. These include consideration of measures that can be taken to minimize the risk of exposure during loading and unloading of the truck as well as during any accidents that might result in spillage of waste from the truck.

The key consideration is use of proper containers. Some type of secondary or outer container was almost always used during transport of infectious waste offsite to hold the plastic bags or other containers of infectious waste, and now an outer container must be used for transport of regulated medical wastes in all states subject to the medical waste tracking regulations.[2] The containers must be sturdy so as to withstand the handling that accompanies loading and unloading of the transport vehicles. Sturdiness is also essential to ensure that the containers can be stacked within the truck without crushing and breaking. The containers are usually sealed or otherwise closed.

Another consideration is that waste transported offsite for treatment must be packaged securely to minimize the chance for spills if there should be an accident

---

*Scrubbers are effective in minimizing emissions of hydrochloric acid from incinerators. Until recently, scrubbers were rarely used on medical waste incinerators as air pollution control devices. Some states now require that all medical waste incinerators be equipped with scrubbers.

while the waste is being transported. Sturdy containers are more likely to survive intact the impact of an accident.

Secondary containers may be single-use or reusable. The single-use (or one-way) container is used to transport waste to the offsite treatment facility where the waste is treated in its container. The reusable container is used to transport the waste to the offsite treatment facility where the container is emptied and the waste is treated. The container is then cleaned and disinfected prior to being returned to the waste generator for reuse. In states subject to the medical waste tracking regulations, reusable containers must have a liner that is not reused, and the containers must be cleaned and disinfected.[2] Easy cleaning and disinfection of reusable containers is therefore essential.

The type of container that is used for waste transport depends on the specifics of the offsite treatment operation. When the waste is to be incinerated, the containers are usually combustible—that is, cartons or fiberboard drums—although reusable containers and even carts are used for some offsite incineration operations. When the waste is to be steam sterilized, the transport containers are usually reusable—for example, drums made of rigid heavy plastic.

REFERENCES

1. U.S. Department of Labor, Occupational Health and Safety Administration. "Occupational Exposure to Bloodborne Pathogens; Proposed Rule and Notice of Hearing." *Federal Register* 54(102):23042–23139 (May 30, 1989).
2. U.S. Environmental Protection Agency. "Standards for the Tracking and Management of Medical Waste; Interim Final Rule and Request for Comments." *Federal Register* 54(56):12326–12395 (March 24, 1989).
3. American Society for Testing and Materials. ASTM Standard #D1709-85: "Standard Test Methods for Impact Resistance of Polyethylene Film by the Free-Falling Dart Method," and ASTM Standard #D4272-85: "Standard Test Method for Impact Resistance for Plastic Film by the Instrumented Dart Drop Method" (Philadelphia: ASTM).
4. Dole, M. "Warning on Autoclavable Bags." *Am. Soc. Microbiol. News* 44(6):283 (1978).

CHAPTER 5

Treatment Considerations and Options

Various options are available for the treatment of infectious wastes. You should evaluate these options during the process of establishing policy and making decisions for the infectious waste management system. Your decisionmaking will include:

1. selection of the place of treatment:
   - onsite at the facility or institution
   - offsite at a commercial treatment facility
   - offsite at another hospital or institution or at a joint treatment facility
2. selection of the type(s) of treatment technology that best suit your needs

In this chapter, the various alternatives are discussed only from the perspective of the decisionmaking process. Technical details and operating features of the different treatment technologies are presented in subsequent chapters. (See Chapter 6 for a discussion of steam sterilization, Chapter 7 for incineration, and Chapter 8 for other treatment technologies.) The process of contracting with a commercial disposal firm is discussed in detail in Chapter 17.

Because of the availability of different options, it might seem relatively easy to find the infectious waste treatment system that is best for a particular institution. However, because of the variety of options and the numerous factors that must be considered, it is difficult to evaluate them all. The discussions in this chapter should help in the decisionmaking process.

Whatever your decisions may be, you will have to be sure that the system is working properly. This is best accomplished through a quality assurance/quality control (QA/QC) program for the waste management system. The final part of this chapter discusses the QA/QC program.

SELECTING THE PLACE OF TREATMENT: ONSITE VS OFFSITE

The first important decision is where to treat the infectious waste. There are two basic options:

- The onsite option provides treatment onsite, that is, at the facility where the infectious waste was generated.
- With the offsite option, the infectious waste is treated at another location, with the untreated waste being transported offsite (that is, away from the facility where the waste was generated).

Onsite Treatment of Infectious Waste

Is Treatment Equipment Available Onsite?

The first prerequisite for the onsite treatment option is the availability of suitable treatment equipment. The available equipment must be able to treat the infectious wastes that are generated at your institution. Capacity must be sufficient to treat the quantities of wastes generated. The treatment and the equipment must comply with all regulatory requirements. (There are relevant environmental regulations* as well as regulations that pertain specifically to infectious waste management.†)

Can Treatment Equipment Be Installed?

If the equipment is not already in place, there might be physical, regulatory, and institutional constraints that would prevent installation of such equipment. Physical constraints are frequently encountered when a hospital wants to build an incinerator; there is just not enough room on the hospital grounds for such construction. Regulatory constraints determine what is feasible from the perspective of regulatory requirements—that is, those that pertain to permitting, waste treatment, and the environmental impacts of the different treatment processes. Institutional constraints reflect the policies and politics of management; for example, the administrators may believe that incineration would be politically unwise in that particular community.

---

*The trend is toward stricter regulatory requirements regarding permissible air emissions from infectious waste incinerators. Some states exclude existing incinerators from new regulatory requirements under a "grandfather clause." In other states, the regulations require that all existing incinerators be retrofitted so that they meet the new regulatory requirements within a specified period of time.
†Many states have promulgated regulations for the management of infectious wastes. At the federal level, medical waste tracking regulations now apply to those states that are included in the two-year demonstration program (June 1989 to June 1991).[1]

What Are the Advantages of Onsite Treatment?

There are several advantages to treating infectious waste onsite. Many infectious waste generators prefer the onsite treatment option because it give them complete control over the waste. They would rather treat the waste onsite than risk potential liability by having it leave the facility untreated. Another advantage is better cost control, because the generator is not subject to the pricing policy of a commercial vendor. In addition, onsite treatment using incineration provides the opportunity for heat recovery, an additional cost benefit. Onsite treatment may reduce the regulatory burden; for example, infectious waste that is treated and destroyed onsite (i.e., incinerated or discharged to the sewer system) is exempt from the tracking requirements of the medical waste tracking regulations.[1] (See Chapter 2 and Appendix A.)

What Are the Disadvantages of Onsite Treatment?

Onsite treatment has several disadvantages. When you elect to treat wastes onsite, you engage in an activity that is not the primary function of your organization (such as providing health care or performing research). Funds must be expended on treatment equipment and operations. You are responsible for meeting all regulatory requirements, which entails keeping current with all new and changing regulations (often a time-consuming and difficult task).

Offsite Treatment of Infectious Waste

What Are the Advantages of Offsite Treatment?

Use of the offsite treatment option allows you to concentrate your efforts on your primary functions. Your concern with infectious waste will be limited to having it properly packaged and picked up for treatment offsite by a reliable vendor. (The offsite treatment facility should be using modern and effective treatment equipment, they should be cognizant of and complying with all regulatory requirements, and they should have an effective quality assurance/quality control program. See Chapter 17 for details on working with commercial treatment/disposal firms.) You should benefit from the economies of scale that a large treatment operation can offer (and that you cannot realize with an onsite operation).

What Are the Disadvantages of Offsite Treatment?

Loss of control over the fate of your waste has been mentioned above. Such loss of control can mean increased liability. You must be sure that the offsite treatment facility is operating in accordance with all relevant regulations; if it is not, you may be liable. You may also be subject to additional regulatory requirements because, under the medical waste tracking regulations (see Chapter 2 and Appendix A), generators must track all infectious waste that leaves the premises untreated.[1]

How Can a Commercial Treatment Firm Be Evaluated?

In brief, you should consider the services offered, the general appearance of the facility and its operations, the dependability and reliability of the firm, its knowledge of and compliance with all regulatory requirements, its concern for the occupational safety and health of its employees, its QA/QC program, its contingency program, its backup plans, and its liability coverage. See Chapter 17 for a detailed discussion of this topic.

What Other Options Are Available for Offsite Treatment?

The commercial treatment facility is not the only option available for offsite treatment of infectious wastes. Some regional facilities and cooperative ventures have been formed to provide shared use of a treatment facility. Two types of shared treatment facilities have evolved. One type is the facility owned by one hospital (or other institution); it treats the infectious waste generated there and also accepts waste from other generators for treatment. The other type is the treatment facility owned and operated jointly by several hospitals or institutions; it treats the infectious waste generated by these institutions and may also accept for treatment infectious waste generated by others. These arrangements are discussed in detail in Chapter 17.

SELECTING THE TREATMENT TECHNOLOGY

Various technologies are effective for treating infectious wastes. These include steam sterilization, incineration, thermal inactivation, chemical treatment, irradiation, and wastewater treatment. Of these, steam sterilization and incineration are used most widely in the United States. Thermal inactivation and chemical treatment are less commonly used but are suitable for certain applications.

No single technology is ideal for all kinds of infectious waste and for all institutions. Each technology has advantages and disadvantages. Therefore, it is not a simple matter to select the technology that is best for you. It may even be advantageous to use more than one treatment technology for the infectious wastes that your institution generates—that is, one technology for some wastes and another for other wastes.

What factors should you consider in selecting the particular treatment technology that will be used at your facility? The factors are numerous, and they include the following:

- regulatory requirements
- ease/difficulty of operation
- need for skilled operators
- applicability
- importance of waste separation
- importance of load standardization
- effect of treatment on the waste
- volume reduction
- occupational hazards
- environmental effects
- costs
- reliability

Regulatory Requirements

The treatment technology that you select must meet all relevant federal, state, and local regulatory requirements. Some regulations require only that the treatment render the waste noninfectious. Other regulations specify which kind(s) of treatment technology must be used and require a waiver for use of an equivalent technology. If you are considering an "alternative technology," you should ascertain what is required to attain such a waiver and if waivers are granted readily.

You may also need permits to construct and operate certain kinds of equipment. Permits are generally not required for installation of steam sterilizers, hammermill/chemical treatment systems, and thermal treatment systems. However, various permits are always required for incinerators, and the permitting process (which includes public hearings) can be long and expensive. (See discussion of incinerator permitting in Chapters 7 and 17.)

In addition, every type of treatment has some impact on the environment, be it on the air, the water, and/or the land. There are regulations to minimize most of these environmental effects. The ease of complying with these regulatory requirements varies with the severity of the potential impact and the stringency of the regulatory requirements.

With steam sterilization, there are practically no relevant regulations. Only the wastewater from the system might be subject to regulatory requirements. It is unlikely that sterilizer wastewater would have any difficulty in meeting the standards for wastewater discharged to the sewer system.

With thermal treatment of liquid wastes, the wastewater must meet the temperature requirements for thermal discharges if the wastewater is discharged into receiving waters such as a river. Incorporation of heat exchangers into the thermal treatment system will reduce the temperature of the discharged wastewater (with energy savings from heat recovery as an added benefit).

With chemical treatment, the chemicals added to the waste change the composition of the wastewater. The chemicals or their reaction products may be present at concentrations that exceed permissible levels.

With the hammermill/chemical treatment system, the principal regulatory concern is the quality of the discharged wastewater. If the final separation system is not functioning efficiently, there could be a significant amount of solid material in the wastewater, with the amount of suspended solids exceeding the permitted limit. The chemical constituents of the wastewater might also be a cause for concern (see above).

With incineration, the most difficult (and expensive) requirement is meeting air pollution control standards that apply to infectious waste incinerators. The regulatory trend is toward emission controls for particulates, acid gases, and organic compounds.* If you are considering incineration as your treatment technology, you must obtain sufficient guarantees from the manufacturer and the contractor that your incinerator—once it is installed and operating—will meet all applicable regulatory requirements. (See Chapter 7 for a more detailed discussion of incineration, including this particular aspect.) If a wet scrubber is used to control emissions, the scrubber water is a potential problem. Even if it is recirculated, it will eventually be discharged to the sewer system and it must then meet the wastewater standards. A final consideration is the solid residue from incineration—bottom ash and flyash. There are concerns that this ash may be hazardous (i.e., by the extraction procedure [EP] toxicity test), which would necessitate its disposal in an RCRA-permitted hazardous waste landfill rather than in a sanitary landfill.

Ease/Difficulty of Operation

The types of equipment used for the various treatment technologies are operated quite differently. Some of the equipment (such as steam sterilizers, hammermills, and equipment for thermal treatment of liquid infectious wastes) may seem simple to operate. Nevertheless, the operators of all types of treatment equipment must be trained in proper operating techniques. They must understand and know how to control the variables in the system (such as load composition and configuration, waste feed rate, and treatment time) to ensure that the treatment cycle is indeed effective. Whatever the treating agent (be it steam, chemical, heat, or radiation), the infectious agents must be exposed to it for sufficient time to ensure treatment effectiveness.

Incineration is much more complex than any of the other treatment technologies. The modern incinerator is a highly complex piece of equipment and therefore difficult to operate correctly. Incorrect operation can cause smoking, incomplete combustion, and other "events" that might lead to ineffective treatment of the infectious waste.

---

*Numerous studies are now being conducted on emissions from infectious waste incinerators. Data from these studies will probably determine the requirements of future regulations.

Need for Skilled Operators

With each kind of infectious waste treatment equipment, there is a need for trained operators. The equipment operator must understand the process as well as the equipment—it is not enough just to load the waste and push a button, because such action could result in incomplete or ineffective treatment as well as occupational exposures and environmental releases.

Highly skilled operators are needed for incinerator operation. This has been true for incinerators built in the last 10 or 15 years, and it is even more important today with newer equipment that often includes air pollution control devices. In fact, some people have termed incineration an art rather than a science. Therefore, the need for and the cost of skilled incinerator operators must be considered when this option is being evaluated.

Applicability

All the treatment technologies can be used for treating at least one type of infectious waste. Some infectious waste is rather uniform in composition (for example, that produced in an industrial or research laboratory), but this is the exception. Most generators of infectious waste (such as health care facilities) generate more than one type of infectious waste. For these generators, it is often important that the treatment equipment be versatile and applicable to various types of waste.

In your evaluation of the various technologies, the following questions are relevant. Will this technology be applicable to most or all of the infectious wastes generated at your institution? Do you have a waste that is best treated in a certain way? Should you select one technology that is most versatile for your types of waste, or should you opt for more than one type of treatment technology?

Certain types of waste are not amenable to treatment by steam sterilization. These include dense items (such as larger masses of tissue or body parts) and large volumes of wastes that are poor conductors of heat (such as containers of animal bedding). See Chapter 6 for a complete discussion of the correlation between waste type and the effectiveness of steam sterilization.

Thermal treatment is most suitable for liquid wastes. In fact, this technology is often ideal for treating infectious liquids and wastewater.

Chemical treatment is a technique rather than a technology. It is especially useful for treatment of infectious liquids.

The hammermill/chemical treatment system is applicable to various infectious wastes. Depending on the model, this equipment can effectively treat sharps and varied laboratory wastes (e.g., petri dishes, glass, tubing, liquids, specimens, and tissue samples). Recent research validated the applicability of this technology to infectious waste streams of mixed composition.[2] Wastes that are unsuitable for this type of processing include substances that are not compatible with the sanitary sewer system, volatile or flammable substances, explosive substances, and heavy metal items.

Incineration is applicable to all types of infectious waste except explosive substances. Certain components of the waste stream, such as plastics and glass, can cause problems. (See "Importance of Waste Separation," below, as well as Chapter 7 for more details.)

Importance of Waste Separation

The infectious waste stream from health care facilities is usually quite variable. It consists of various types of infectious waste, comprised of body tissues, liquids, cloth, paper, metal, plastics, rubber, and glass. This variability in the waste stream can create some problems when certain technologies are used to treat the waste, and the removal of certain items from the waste stream may be necessary or desirable before treatment.

It is essential to note that, because of occupational safety considerations, the separation of certain items from the infectious waste stream must be done as source separation—that is, when the items are discarded, the treatable and untreatable infectious wastes must be discarded directly into separate containers. The SOPs must specify the techniques for such separation. Sorting through the waste to remove untreatable items is *not* an acceptable procedure.

Certain wastes are not amenable to steam sterilization. (See "Applicability," above, and Chapter 6.) Some separation of certain components of the infectious waste stream may be necessary to exclude untreatable wastes from those that will be steam sterilized.

Incineration is the technology that can effectively treat most infectious wastes. However, incineration of liquids is inefficient and energy-intensive, and other treatment of these wastes is often preferable. Plastics in the waste can cause problems such as temperature spikes in the incinerator, corrosion, and hydrogen chloride emissions. The presence of glass in the waste could also cause problems (such as slagging) when the waste is incinerated. (See Chapter 7 for a complete discussion of wastes that are not compatible with incineration.) Because of concerns about incinerator durability and downtime, it might be desirable to minimize the amounts of plastic and glass in the waste stream that are incinerated.

The hammermill/chemical treatment process has some limitations because it is not suitable for treating all infectious wastes. (See "Applicability," above). Therefore, some source separation is necessary to exclude certain wastes from the waste stream that is fed to the unit for processing.

Importance of Load Standardization

With steam sterilization, load standardization is essential to ensure effective treatment. This means standardizing waste type and the configuration of the load within the sterilizer. The required operating conditions for each standard load must be determined and standardized, then incorporated into the SOPs for operating the sterilizer. This process can be simplified by establishing a single worst-case load, determining the operating conditions needed for that load, and standardizing them into one set of SOPs for operation of the steam sterilizer. These SOPs are then used to treat each load of infectious waste that is steam sterilized.

With incineration, load standardization may be desirable theoretically, but it would be impractical to implement during routine incineration operations because the content of the infectious waste stream is so variable (regarding composition, heat content, moisture content, etc.). Nevertheless, the incinerator must always be operated properly to ensure complete combustion of all the wastes.

With hammermill/chemical treatment, load standardization can be useful because the concentration of disinfectant can be varied with the type of waste being treated. Some equipment models are designed to process only sharps. Other models can process a stream of mixed infectious wastes; in this application, it may be necessary to alternate the types of waste being fed to the hammermill (i.e., to alternate soft items such as gloves, gauze, and tubing with hard items such as glass and sharps) in order to achieve maximum processing effectiveness and efficiency.[3]

Effect of Treatment on the Waste

You should be concerned about the effect that treatment has on the waste besides rendering it noninfectious. What is the appearance of the waste after it has been treated? Does it look the same? Is it recognizable as something that was once infectious waste? Is it apparent that the waste was treated? Can the generating facility be identified through papers and labels? Are needles and syringes reusable?

With steam sterilization, the waste remains intact and recognizable. Steam treatment wets the waste and removes contained air, but otherwise the waste is the same as before treatment. Any papers and labels remain intact. Needles and syringes remain intact and reusable.

With incineration, the waste is destroyed by burning. Under proper operating conditions, burndown is complete and only incombustible material remains in the ash. Papers and labels will be completely burned. Disposable needles will be friable and flaky and cannot be reused.

In the hammermill/chemical treatment process, the waste is shredded and ground so that only small pieces of waste remain after treatment. Some pieces of waste may recognizable (as, for example, part of a needle or a piece of plastic), but the waste seldom remains intact as it passes through the hammermill. (One exception is small rubber stoppers.) With proper operation of the hammermill, no paper, needles, or syringes will be intact after treatment.

Volume Reduction

Volume reduction is important because all solid residues from treatment are eventually disposed of in landfills. Landfill tipping fees have been increasing as landfill capacity is used up and landfills must be sited farther from urban areas. Tipping fees are usually based on the volume of the waste, although there is now a strong trend toward basing tipping fees on weight rather than volume.[4]

Therefore, it is important to reduce the volume—and the weight—of the waste residues that will be landfilled. The greater the volume reduction, the less material that has to be landfilled, and the lower your final disposal costs.

Steam sterilization results in a volume reduction of about 30%. Although the waste is not really changed by the steam, air in the waste is removed when the sterilizer chamber is evacuated as the first step in the treatment process, and this results in a volume reduction.

The hammermill/chemical treatment process results in some volume reduction because the process eliminates air and liquids from the waste. In addition, some suspended solids are discharged in the wastewater to the sewer system, with the amount of pulp and suspended solids in the wastewater depending on the separation efficiency at the end of the process train. A volume reduction of up to 8:1 has been reported for this technology when general infectious wastes are treated.[2]

The greatest volume reduction is achieved with incineration. The ash residue from incineration usually has a volume of only 5–15% that of the original waste.

Occupational Hazards

Occupational hazards must be of concern to everyone interested in reducing risk and liability. Risks are inherent in the handling of infectious wastes. (For details, see Chapter 14.) All operators of infectious waste treatment equipment incur the same risks as do other infectious waste handlers. OSHA regulations for the prevention of occupational exposure to blood-borne pathogens* should greatly reduce the potential for this type of exposure among all handlers of infectious waste, including treatment equipment operators.

Operators of infectious waste treatment equipment also incur risks that are specific to each type of equipment and its operation. Only those risks that are related to operation of infectious waste treatment equipment are discussed in this section.

With steam sterilization, the principal occupational hazard relates to opening the sterilizer door at the end of the cycle. If the cooldown stage is not complete, there is danger of burns from the steam. Another hazard is exposure to vapors generated during the sterilization process that may be released when the door is opened. These hazards can be reduced by engineering controls (e.g., door interlocks and ventilation hoods and ducts), by the use of appropriate personal protective equipment (such as moisture- and heat-resistant gloves), and by adoption of and adherence to appropriate SOPs for equipment operation.

---

*At the time this book was published, the regulations were still in the proposal stage, and final regulations had not yet been promulgated. For the proposed rule, see Reference 5.

With the hammermill/chemical treatment system, the potential occupational hazards are exposure to the treatment chemical, injury from moving parts (hammermill and conveyor belts), and exposure to a high noise level. These hazards can be minimized by proper equipment design (such as guards to prevent exposure to moving parts), by the use of appropriate personal protective equipment (such as—when necessary—hearing protection, latex and heavy gloves, aprons, shoe covers, and face shields or masks and safety goggles), and by the implementation of suitable SOPs for equipment operation.

With incineration, the occupational hazards are greater because of the nature of the incinerator. Auxiliary equipment often includes conveyor belts, rams, hatch doors, etc.—each with inherent dangers if the design is bad or the equipment is misused. In addition, there are the hazards inherent in fire and high temperatures. As with the other technologies, risks to equipment operators can be reduced by engineering controls, use of personal protective equipment, and institution of appropriate SOPs.

Environmental Effects

Each type of treatment technology has some impact on the environment. There are air emissions from incinerators. Wastewater is produced by steam sterilizers, by incinerators with wet scrubbers, by chemical treatment, and by thermal treatment. Almost all the technologies produce solid residues that must be disposed of in landfills: the treated waste from steam sterilization, fly ash and bottom ash from incineration, and the solid component resulting from other treatments of solid infectious waste. (See the section entitled "Regulatory Requirements" in this chapter for a discussion of the environmental impacts of each treatment technology.)

Costs

The costs of treatment involve capital costs for the treatment equipment, operating costs (including labor, materials, and utilities), and maintenance costs. These costs vary, depending on the type of equipment.

The incinerator is much more expensive than any of the other types of equipment that are used for treating infectious wastes. Capital costs for an incinerator are the highest, and the addition of air pollution control equipment usually doubles the capital cost. Incinerator operation requires the full gamut of utilities (electricity, water supply, and sewer) and an auxiliary fuel. Incinerators should be operated only by trained and skilled operators, and, therefore, labor costs are relatively high. Maintenance is also expensive: the refractory needs constant maintenance as well as periodic replacement.

Steam sterilizers are much less expensive than incinerators. Sterilizers require a steam line, electricity, a vent, and a sewer connection. They should be operated by trained personnel, but highly skilled operators are not needed. The operation is not labor-intensive, and one trained person can operate more than one sterilizer. Maintenance costs are not too high; routine maintenance involves gasket replacement, etc. The sterilizer has no moving parts other than the door and the drain valve.

Equipment for the hammermill/chemical treatment system is rather expensive (less than the cost of an incinerator but more than that for a steam sterilizer). Utility needs are water, electricity, an outside vent, and a sewer connection. Materials needed include the treatment chemical and (for the large unit) a substance to clean the hammermill to prevent corrosion. Routine maintenance of the equipment and parts replacement are necessary. The hammer requires routine maintenance and periodic replacement of the blades. The hammermill housing has to be rebuilt or replaced periodically. In the large units, the conveyor belts for feeding the waste and draining the final slurry also need some routine maintenance and periodic replacement.

Reliability

You want the infectious waste treatment system that you select to provide reliable service. You generate infectious wastes on a routine basis, and you need treatment equipment that will operate routinely, without too many problems. Downtime is expected. The question is: how will this interfere with your operations?

There are two types of downtime on equipment. One is the scheduled downtime for the routine maintenance and parts replacement that all equipment requires. The other is the nonscheduled downtime that results from incidents such as catastrophic failure of the equipment. Inasmuch as such failures are generally not anticipated, replacement parts are usually not available and long downtime can ensue.

For the steam sterilizer, routine maintenance activities are generally relatively simple and quick. Downtime for routine maintenance operations is usually brief. With the hammermill, maintenance procedures are more complicated, especially when the hammer or the hammermill housing has to be removed for replacement. Downtime is usually longer than that required to maintain and replace parts on steam sterilizers. Medical SafeTEC estimates that the downtime for their equipment averages 5% of operating time.[6]

Maintenance operations are most complicated with the incinerator. The technology is sophisticated, and downtime can occur fairly frequently. Replacement of the refractory usually requires a couple of weeks.

Summary

The information presented in this section cannot be summarized as a recommendation for the use of a specific type of treatment technology. The selection of one or more treatment technologies is an individual decision that must be based on the particular variables, constraints, and other factors that prevail at each institution.

Nevertheless, the information has been tabulated in an attempt to make comparison of the different technologies somewhat easier. Because of changing technology and the impossibility of quantifying many factors, comparisons are presented in relative rather than absolute terms. (See Table 5.1.)

Table 5.1. Comparison of Treatment Technologies

	Type of Treatment Technology		
Factor	Steam Sterilization	Incineration	Hammermill/ Chemical Treatment
Operations			
Applicability	Most infectious wastes	Almost all infectious wastes	Most infectious wastes
Equipment operation	Easy	Complex	Moderately complex
Operator requirements	Trained	Highly skilled	Well trained
Need for waste separation	To eliminate non-treatable wastes	None	To eliminate nontreatables; for proper feeding
Need for load standardization	Yes	No	When feeding by type of waste
Effect of treatment	Appearance of waste unchanged	Waste burned	Waste shredded and ground
Volume reduction	30%	85–95%	Up to 85%
Occupational hazards	Low	Moderate	Moderate
Testing	Easy, inexpensive	Complex, expensive	Protocol under development
Potential side benefits	None	Energy recovery	Use of effluent in laundry
Onsite/offsite location	Both	Both	Both
Regulatory Requirements			
Medical waste tracking regulations	Applicable	Recordkeeping	Not applicable
Applicable environmental regulations	Wastewater	Air emissions, ash disposal, wastewater	Wastewater
Releases to air	Low risk via vent	High risk via emissions	Low risk via vent

continued

Table 5.1. Continued

	Type of Treatment Technology		
Factor	Steam Sterilization	Incineration	Hammermill/ Chemical Treatment
Releases to water	Low risk via drain	Low risk via scrubber water	Moderate risk via wastewater
Disposal of residue	To sanitary landfill; potential problem with *red* bags	Ash may be a hazardous waste; if so, to RCRA-permitted landfill	Effluent to sanitary sewer; residue to sanitary landfill
Permitting requirements	None	For siting, air emissions	None
Costs			
Capital costs	Low	High	Moderate
Labor costs	Low	High	Moderate
Operating costs	Low	High	Moderate
Maintenance costs	Low	High	Moderate
Downtime	Low	High	Moderate to high

ESTABLISHING A QUALITY ASSURANCE/QUALITY CONTROL PROGRAM

A QA/QC program is essential to the infectious waste management system to ensure that established procedures are being followed. You should establish the QA/QC program in accordance with the details of the plan for infectious waste management. The program has three phases: development of the policies and procedures, their implementation, and verification.

Important elements in the QA/QC program can be incorporated into the infectious waste management procedures. For example, the procedures could include the following:

- Only waste that is properly packaged in the designated containers will be picked up from the nursing stations, laboratories, and other places where infectious wastes are generated.
- Only designated types of containers are loaded into steam sterilizers, incinerators, or other types of treatment equipment.
- Incinerator operation is stopped when smoke is emitted from the stack.
- Incinerator operation is stopped when the quality of the ash indicates incomplete burnout of the waste.
- Only waste that has obviously been treated is placed in dumpsters for removal from the facility.

With a QA/QC program in place, there is reasonable assurance that infectious wastes are being handled and treated properly, and that only treated waste is being sent to a landfill for disposal. Nevertheless, the existence of policies and procedures does not ensure that they are being followed.

When the waste management and QA/QC policies and procedures are in place, it is essential to conduct spot checks to verify that policies are being implemented and that the established SOPs of the waste management and QA/QC programs are indeed being followed. The QA/QC program should include checks at various steps in the waste management process (e.g., discard, collection, storage, and treatment operations) to ensure that all procedures are correctly followed and that all infectious waste reaching each of the check points has been handled properly up to that point.

Another aspect of the QA/QC program is a system for periodic checking of the various elements of the waste management plan. This part of the QA/QC program involves recordkeeping and review of the data and is useful for fine-tuning the waste management plan and for providing documentation of activities. For this part of the QA/QC program, it is important to collect routinely and to review periodically data for the following:

- quantities of infectious waste collected at the various locations (significant changes may indicate that infectious waste is bypassing the system or that noninfectious waste is being mixed with the infectious wastes)
- frequency of cleaning and disinfection of collection carts
- quantities of infectious waste placed in storage, with a written record for each container that includes dates of waste generation, placement in storage, removal from storage, and treatment or shipment offsite
- use of treatment equipment with operations schedule, maintenance and repair work, and automatic recording data (for example, temperature graphs from steam sterilizers)
- tests of equipment operations and treatment effectiveness (for example, biological indicator tests of steam sterilizers, tests of incinerator stack gases)
- quantities of wastes treated by each unit of treatment equipment
- quantities of treatment residues hauled away for landfill disposal
- quantities of infectious wastes hauled away for treatment at an offsite facility

When a good infectious waste management plan and a good program for quality assurance/quality control have been implemented, an infectious waste generator can be reasonably certain that the system is working as intended. Infectious wastes will be handled properly and treated in accordance with established procedures. As a result, only treated waste will be sent to the landfill for disposal or discharged to the sewer system.

For the hospital or commercial treatment facility that treats infectious waste generated by others, the QA/QC program has another dimension because the treatment facility can be held liable for incidents arising from the waste it treats. Therefore, treatment facilities should insist that the infectious waste generators have good QA/QC programs in effect and that the QA/QC procedures are being implemented.

REFERENCES

1. U.S. Environmental Protection Agency. "Standards for the Tracking and Management of Medical Waste; Interim Final Rule and Request for Comments." *Federal Register* 54(56):12326–12395 (March 24, 1989).
2. Denys, G. A. "Microbiological Evaluation of the Medical SafeTEC Mechanical/ Chemical Infectious Waste Disposal System." Paper presented at the 89th Annual Meeting of the American Society for Microbiology, New Orleans (May 14–18, 1989). *Program of the 89th Annual Meeting of ASM*, p. 101, Abstract #Q 57.
3. Medical SafeTEC, Inc. *Infectious Waste Disposal System, Model Z-5000HC, Operations Manual* (Indianapolis, IN: Medical SafeTEC, Inc., undated), p. 12.
4. O'Leary, P. R. Assistant Professor and Assistant Chairman, Department of Engineering Professional Development, University of Wisconsin, Madison, WI. Personal communication to Judith G. Gordon (March 29, 1989).
5. U.S. Department of Labor. Occupational Safety and Health Administration. "Occupational Exposure to Bloodborne Pathogens; Proposed Rule and Notice of Hearing," *Federal Register* 54(102):23042–23139 (May 30, 1989).
6. Wilson, J. H. Vice President, Medical SafeTEC, Indianapolis, IN. Personal communication to Judith G. Gordon (June 27, 1989).

CHAPTER 6

Steam Sterilization of Infectious Waste

For many decades steam has been used to sterilize medical supplies, equipment, and instruments.[1] Steam sterilization has also been proven to be a reliable way to treat infectious wastes.[2-4] As of 1988, 27 states recommended steam sterilization as a treatment method for infectious waste.[5] In fact, steam sterilization is considered to be the preferred treatment method for certain types of infectious waste.[6,7]

CONDITIONS REQUIRED FOR STEAM STERILIZATION

Steam sterilization requires that the infectious agents in the waste be exposed to a high temperature for a sufficiently long duration. However, it is incorrect to think of steam sterilization as a simple enhancement of dry heat sterilization or that the primary role of steam in steam sterilization is to facilitate heating of the material or waste. Steam sterilization relies on the fact that saturated steam is a powerful sterilizing agent on its own. In addition to the sterilization criteria of time and temperature, a third criterion must be met for steam sterilization: direct contact of saturated steam with the infectious agents.

Table 6.1 describes the time and temperature requirements for steam sterilization *under ideal conditions,* that is, when there are no interferences in heating the waste or attaining direct contact between the infectious agents and saturated steam. When comparing identical temperatures, steam sterilization is 30 to 200 times faster than dry heat sterilization. This attests to the sterilizing properties of steam. For steam sterilization of infectious waste, time and temperature conditions to ensure sterilization must be determined through testing, as explained below.

Table 6.1.   Time and Temperature Requirements for Steam Sterilization[a]

Temperature		Spore Kill Time[b]
(°F)	(°C)	(minutes)
240	116	30
245	118	18
250	121	12
257	125	8
270	132	2
280	138	0.8

[a]Data from E. Hanel, Jr., "Chemical Disinfection," in *Control of Biohazards in the Research Laboratory,* Course Manual (Baltimore, MD: Johns Hopkins University, School of Hygiene and Public Health, 1981).
[b]In steam sterilization, exposure time for treatment is usually at least double the kill time.

Saturated Steam and Steam Production

Steam serves a dual role in steam sterilization: steam augments the properties of heat that destroy infectious agents and facilitates the transfer of heat to the waste. Steam is water in a gaseous physical state. For efficient sterilization, the steam must be saturated. Saturated steam is water vapor just on the gas side of the gas/liquid phase boundary; when water is gradually heated beyond its boiling point, it first becomes saturated steam. Saturated steam typically contains some very small water droplets held in suspension or entrained in the steam, although boiler operators attempt to keep the percentage of free moisture in the steam less than 3%.

In practice, most waste either is moist or wet or contains water. When water is inherently present in the waste, steam can be generated within the waste load *if* the waste is thoroughly heated to sufficient temperatures. Production of saturated steam within the waste load also requires an adequate level of moisture in the waste. Waste-generated steam is a common occurrence. For example, although a closed petri plate prevents outside steam from entering it, moisture in the medium within the plate can, if heated sufficiently, generate steam to destroy the infectious agents on the medium.

If heat is added to saturated steam beyond the point at which all of its free moisture is removed, the steam is called superheated. Superheated steam is not useful for sterilization and is to be avoided. This is usually not a problem because boilers, heating systems, and steam sterilizers that use steam are designed to produce and use saturated steam.

Types and Features of Steam Sterilizers

Two types of devices are commercially available for steam sterilization: autoclaves and retorts. Both have a chamber constructed to withstand the temperature and pressure of steam sterilization. At temperatures practical for steam sterilization, steam is pressurized to about 15 to 30 psi gauge.* Chamber sizes range from one to several thousand gallons. The chamber has an inlet for steam and an outlet for steam, air, and condensation.

An autoclave is the most common type of steam sterilizer. An autoclave is characterized by a having a steam jacket that surrounds the pressure chamber (see Figure 6.1). The jacket is filled with steam prior to loading. Once the autoclave is loaded

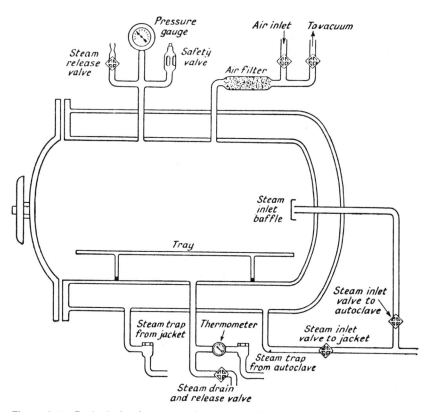

Figure 6.1. Basic design for an autoclave operated from a steam main.

and closed, steam enters the chamber. Since the walls of the chamber are pre-heated, the warmup period for the waste is faster. Preheating of the chamber and its contents also reduces condensation of water inside the chamber and on the waste. Autoclaves can be further classified into two types depending on the method used to remove air from the chamber: gravity displacement (more common) and

---

*Pounds per square inch, gauge pressure: read directly from the pressure gauge. At sea level, absolute pressure is 14.7 psi more than gauge pressure.

pre-vacuum. In gravity displacement models, lighter steam is fed into the chamber to displace the heavier air. Pre-vacuum autoclaves have vacuum pumps that remove air from the chamber prior to the introduction of steam. For steam sterilization of wastes, pre-vacuum models offer a significant advantage in efficiency over gravity displacement models.

Retorts differ from autoclaves in that they don't have a steam jacket. Additional heat from the steam is necessary to heat the walls of the pressure chamber. Condensation is more likely to form when there is no preheating jacket, so the material being sterilized usually gets wet. Wetting is usually not as important a consideration with wastes as it is with reusable items. For faster and more efficient operation, some retorts are designed to routinely operate at a higher pressure and temperature than what is commonly used for autoclaves; this is a great advantage for treating infectious waste, as explained below. Retorts sold today are typically very large units designed to steam sterilize large volumes. These large units are used by commercial infectious waste management firms or at regional waste facilities.

Operation of a Steam Sterilizer

A typical processing cycle for steam sterilization of infectious waste includes six operational stages.

Loading

Waste containers, such as bags or needle boxes, are placed on racks or in carts (i.e., load carriers) for easy handling and loading. When loaded into the chamber, waste containers and load carriers must be carefully arranged and configured to optimize the removal of air. Chemical or biological indicators are also placed in the waste load to monitor the sterilization process. After waste to be sterilized is put in the pressure chamber, the door is closed and sealed.

Pre-vacuum

In steam sterilizers with a pre-vacuum feature, a vacuum is then drawn upon the chamber to remove air. Some pre-vacuum autoclaves can vary the duration and number of vacuum cycles and the depth of evacuation.

Steam Charging

This stage is when saturated steam is fed into the chamber from a boiler or other steam source. The remaining air, which is heavier than steam, drains from the bottom of the chamber as it is displaced by the lighter steam. In gravity displacement models, displacement of air by steam is the only means of air removal, which is a critical step in steam sterilization. This stage lasts only two to five minutes. The steam charging stage also commences waste heating to raise the waste temperature to that required for sterilization.

Steam Sterilization

The steam sterilization stage begins on attainment of the designated sterilization temperature at the temperature controller. During this stage the chamber temperature and pressure are held as stable as possible until a sufficiently long time has elapsed to ensure sterilization. In contrast to the stable chamber temperature, waste temperatures typically climb during the first half of this stage. Waste temperatures throughout the load equilibrate at or near the chamber temperature near the end of this stage. Additional steam may be automatically fed into the chamber during this stage as necessary to compensate for heat losses and steam condensing inside the chamber.

Steam Discharge

After the predetermined exposure period has elapsed, steam is then vented from the chamber. This lowers the chamber temperature and equilibrates the chamber pressure to atmospheric pressure.

Unloading and Disposal

As part of the quality assurance procedures described below, the results of chemical indicators must be noted during unloading. Disposal of steam sterilized waste is also discussed below.

Exposure Period

In the operation of most steam sterilizers, the steam charging and sterilization stages are considered together as one processing or exposure period. The critical parameters of time and temperature must be satisfied during this period, exclusive of the pre-vacuum and steam discharge stages. As described below, the appropriate duration of the exposure period is determined through the development of the standard operating procedures.

Instrumentation and Control

Steam sterilizers differ greatly in their degree of instrumentation and automatic controls. A temperature control bulb in the sterilizer's drain line usually serves as the principal temperature monitor and chamber temperature controller. The drain is an important point for temperature measurement; it is where the cooler air leaves the chamber as it is displaced by steam. Under ideal conditions, the drain is where the lowest chamber temperature is expected. Attainment of the sterilization temperature at the drain implies that air has been removed from the chamber.

Several additional devices may be used to measure, monitor, and record the temperature inside the chamber and within the waste load itself. Sophisticated autoclaves have thermocouples wired to a potentiometer with a digital readout and/or chart recorder. These sensors continuously plot the waste load temperatures during the exposure period. Thermocouples can be placed at various points in the waste load to monitor differences in load temperatures. This is a very useful feature for steam sterilization of infectious waste. The plots (time/temperature profiles) can reveal heating problems due to load configuration or differences in waste load characteristics.

The pressure and vacuum depth of the chamber are monitored by one or more gauges. Other automated features include timers to regulate the duration of operational stages, controllers to ensure that sterilization temperatures are maintained for the specified time during the exposure period, and programs that automate several operational parameters.

The one critical parameter that cannot be measured is direct steam contact. This is important to remember because a mixture of steam and air can have the same temperature and pressure as saturated steam. When such mixtures exist in the chamber, temperature measurements may incorrectly indicate that the air has been completely removed and that sterilization conditions have been achieved. Temperature measurements, especially when thermocouples are placed directly in the waste load, may imply good steam penetration, but operators must understand that such measurements can be misleading when air is not removed.

POTENTIAL PROBLEMS IN INFECTIOUS WASTE STERILIZATION

The three critical parameters for steam sterilization are time, temperature, and direct steam contact. Once the waste attains the proper temperature and saturated steam is given the opportunity to come into direct contact with the infectious agents, assuring a sufficiently long sterilization time is relatively straightforward. However, there are many factors that can interfere with direct steam contact or attainment of sterilization temperatures (see Table 6.2). These interfering factors are often present while steam sterilizing infectious waste. As a result, the time and temperature requirements specified for ideal conditions (Table 6.1) are insufficient for infectious waste. This section describes the factors that can interfere with steam sterilization, and potential solutions are discussed below.

**Table 6.2.   Summary of Potential Problems and Interferences with Steam Steriliza-
tion and Solutions and Optimal Conditions**

Potential Problems and Interferences

Interferences with Waste Heating
- Excessive mass or weight
- Low heat capacity
- Low heat conductivity
- Barriers to heat transfer

Interferences with Direct Steam Contact
- Incomplete air removal
- Waste containment
- Inappropriate waste types
- Poor steam quality

Solutions and Optimal Conditions

Appropriate Waste Types

Appropriate Waste Containment, Load Carriers, and Load Configurations
- Alternative bags and other primary waste containment
- Alternative load carriers
- Alternative load size and configuration

Alternative Conditions
- Increasing the depth or number of vacuum cycles
- Lengthening the exposure period
- Increasing the exposure period temperature

Other Procedures
- Opening the bags or containers[a]
- Punching holes in the bags[a]
- Adding water[a]

[a]This procedure may pose a risk to waste handlers. Attempt only after other
alternatives have been pursued. See text for precautions.

Factors That Interfere with Waste Heating

Steam sterilization can be unsuccessful if waste or waste load characteristics
interfere with heating and attainment of sterilizing temperatures throughout the
waste load. Potential interferences are listed below.

Excessive Mass or Weight

Heavy or massive waste requires relatively more heat to reach sterilization temperatures. Heat is absorbed by a waste in proportion to a waste's mass or weight.
Heating 20 lb of infectious waste to the necessary temperature requires twice as

much heat as 10 lb. Given a constant heat source, very heavy waste may require an extended exposure period to allow the waste to reach the required temperature.

Low Heat Capacity

Waste types differ in the amount of heat needed to reach sterilization temperatures even when their weights are identical. The amount of heat needed to incrementally raise the temperature of a material is called its heat capacity. Waste with a low heat capacity insulates itself from thorough heating.

Low Heat Conductivity

Heat conductivity, the movement of heat through an object, is an important way that heat is transferred throughout the waste. A waste that is a poor heat conductor will take longer to reach sterilization temperatures. For example, plastic is a poor conductor of heat; therefore plastic containers heat more slowly than containers made of metal, which is a good heat conductor.

Barriers to Heat Transfer

Steam is an excellent heat transfer medium, working primarily through convection and condensation of steam on the waste. Anything that prevents steam from reaching the waste or penetrating the waste load is a barrier to heat transfer. The efficiency of heat transfer also depends on the amount of available surface area of the waste. Wastes that have a small ratio of surface area to volume also heat poorly.

Factors that Interfere with Direct Steam Contact

Steam sterilization requires direct contact between saturated steam and the infectious agents. Several factors that can interfere with this requirement are discussed below.

Incomplete Air Removal

If air is not removed from the chamber it can prevent steam from directly contacting the waste. Alternatively, air that remains in the chamber can mix with the steam so that the steam is no longer saturated. In either case, the waste is not heated efficiently and saturated steam is prevented from directly contacting the infectious agents. Air can be difficult to remove. Containers can impede air removal as well as prevent steam penetration. Air can pool and pocket in low areas when overpacks and load configurations prevent its displacement by steam. Figure 6.2 plots waste temperatures of a load in which the air has not been completely removed in comparison with a load in which the air has been properly

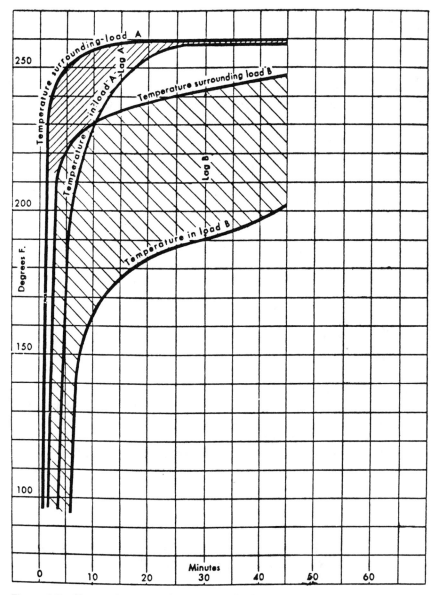

Figure 6.2. Temperature curves for steam sterilization with and without complete removal of air. *Run A:* When air was completely discharged from the chamber, temperature in the package rapidly approached that of the surrounding steam. *Run B:* When only a small amount of air was discharged, temperature in the package lagged about 50°F behind that of surrounding steam throughout the 45-minute exposure. Pressure was maintained at 20 psi. Loads A and B were identical. (From G. Sykes, *Disinfection and Sterilization,* copyright 1958 by Van Nostrand Reinhold, New York. Reprinted with permission.)

removed, demonstrating how incomplete air removal interferes with waste heating as well.

Waste Containment

Waste containers are the most common impediment to direct steam contact. The waste container must allow steam penetration or entry. Container materials and their method of closure are critical factors. Although plastic bags do a good job of safely containing waste for handling, many plastics do not readily allow steam to penetrate and directly contact the contaminated waste. Some plastic bags that have been sold as being "autoclavable" have been shown to be poor at allowing steam penetration.[8,9] Similarly, a tightly closed bag or a needle box that has been sealed with tape protects waste handlers but allows little opportunity for steam to enter. Waste containers that are impervious to steam (metal, glass, thick plastic, etc.) and are sealed or tightly capped have no chance of steam sterilization unless the waste contains a significant amount of water to generate steam internally. Some containers and wastes melt when exposed to sterilizing temperatures. Thermal instability can interfere with steam sterilization when melted plastic encapsulates wastes and prevents steam penetration (e.g., disposable petri dishes), but it can also facilitate direct steam contact if the bag decomposes. When infectious waste is to be steam sterilized the choice of waste containment is critical.

Inappropriate Waste Types

Waste oils and greases that repel water and do not allow direct contact between the infectious agents and saturated steam cannot be steam sterilized. Large animal carcasses, which are difficult to heat and do not allow steam penetration, may also be considered an inappropriate waste for steam sterilization. There are other wastes that are unsuitable for steam sterilization, either because their inherent interferences are too difficult to overcome or because other waste treatments are easier or more appropriate.

Poor Steam Quality

Steam for sterilization is usually produced specifically for the steam sterilizer or autoclave or is connected to the steam lines of a boiler or general heating system. It is important that saturated steam is supplied.

PURSUING OPTIMAL CONDITIONS FOR STEAM STERILIZATION

Interferences with waste heating and direct steam contact can usually be overcome through the use of more rigorous sterilizing conditions, appropriate containment and load carriers, and adherence to standard operating procedures that have been proven to ensure sterilization. Note that these conditions will likely be entirely different and more stringent than those used for routine sterilization of reusable articles and supplies. As discussed in the following sections, sterilization can only be assured when standard operating procedures are established, which usually requires a trial-and-error investigation in an effort to identify optimal steam sterilization conditions. Employee training and an ongoing quality assurance program for steam sterilizer operation are also necessary. Knowledge of the potential solutions for steam sterilization problems described in this section will make those tasks easier.

Appropriate Waste Types

Any waste that allows steam to easily penetrate or is inherently moist is appropriate for steam sterilization. The National Research Council and U.S. EPA recommend steam sterilization as the method of choice for treating waste cultures, stocks of infectious agents, associated biologicals, and contaminated labware.[6,7] Further, the EPA recommends steam sterilization or incineration for isolation wastes, blood and blood products, and contaminated sharps. EPA also notes that while steam sterilization may be useful for pathological wastes and animal carcasses, for aesthetic reasons these wastes should be incinerated or otherwise rendered unrecognizable before final disposal.[10]

Appropriate Waste Containment, Load Carriers, and Load Configuration

Selecting optimal containers, carriers, and configuration requires knowledge of supplied materials and operator practices and consideration of the factors that can interfere with steam sterilization.

Bags and Other Primary Waste Containment

Plastic bags, composite paper-plastic bags, boxes, and bins are all used as primary containers for infectious waste. Thickness, construction, and type of material have all been shown to affect steam penetration through waste containers. Paper-plastic composite bags work well in some situations (supplied as a paper bag with a thin polyethylene liner), whereas problems have been reported with plastic bags.[8,9] Note that the term "autoclavable" may only refer to a material's thermal stability at sterilization temperatures (meaning it won't melt) and generally does not assure that steam can penetrate it. Most types of plastic, if thick enough, can completely block steam penetration.

Comparing similar construction and thickness, polyethylene appears to be superior to polypropylene for steam sterilization, but it is not as sturdy for waste handling.[2,4] Polypropylene bags have been shown to interfere with steam penetration and some suppliers recommend that their tops be loosely constricted when used for steam sterilization. Some autoclave bags are constructed so that their closure leaves a small orifice (e.g., bags with elastomeric closures) to provide a pathway for steam to enter. Alternatively, bag openings can be loosely constricted using a twist tie.[2]

Double bagging nearly always impedes steam penetration. However, a recent report recommended the use of meltable polyethylene plastic bags inside heat-stable polypropylene. Just before the waste is to be sterilized, the exterior bag is opened and folded down. At sterilization temperatures, the interior bag melts and allows direct steam contact, while the exterior polypropylene bag contains the waste. After processing, the exterior bag is gathered around the waste for transport and disposal.[4]

Because materials and their construction vary by manufacturer and even by manufacturer's lot, testing (as described below) is the only way to ensure that containers do not unduly interfere with steam sterilization. It is a good idea to specify in the institutional purchasing contract that bags furnished by suppliers must minimize interferences with steam sterilization.

Load Carriers

Ideally, load carriers, trays, and carts should be designed to allow free ventilation and not have cavities where air can pool. When possible, suspend the waste above the floor with a perforated grate or rack to allow steam to escape. Rutala et al. found that trays used to hold waste bags made of stainless steel were significantly more efficient in transferring heat than polypropylene trays, which are made of an inferior heat conductor.[2] Buckets, high-walled trays, and other deep load carriers can interfere with the drainage of air, and they should be avoided. For some wastes, however, trays, tubs, or some other type of secondary containment are necessary to contain routine or periodic spillage. For example, disposable petri plates and other plastic wastes often melt at sterilization temperatures. Liquified agar and other liquids can soil the autoclave floor and obstruct the drain if uncontained. When secondary containment is necessary, use oversized trays with low lips to facilitate drainage and removal of air.

Load Size and Configuration

Rutala et al. found that waste heating was more efficient with smaller waste loads.[2] In many cases two small loads will take less time and more easily achieve consistent sterilization than one large load. Loads should also be configured so that air can easily be displaced by steam. Tightly stacked waste containers do not allow ventilation. Placing bags on multi-level racks will result in much more efficient sterilization than piling them in a cart.

Alternative Conditions

The use of alternative conditions is the most common means of overcoming interferences with steam sterilization.

Increasing the Depth or Number of Vacuum Cycles

Variations in the pre-vacuum stage that can facilitate air removal include increasing the number of vacuum cycles, lengthening the duration of the pre-vacuum stage and, on occasion, increasing vacuum depth. A more rigorous pre-vacuum stage is especially effective for wastes, containers, or load carriers that interfere with air displacement.

Lengthening the Duration of the Exposure Period

The duration of the exposure period (usually considered to be the combined steam charging and sterilization stages) is typically extended for infectious waste. This can usually overcome most interferences. Prolonging the exposure period provides more opportunity for steam penetration and heat transfer to thoroughly heat the waste load. For routine sterilization of supplies, institutions commonly use an exposure period that is twice the duration of the spore kill time. For infectious waste, however, the exposure period may be up to seven times as long as the spore kill time. For example, it is not uncommon to extend the exposure period for infectious waste to 30, 60, or even 90 minutes at 250°F. Prolonging the exposure period is the most common way to achieve sterilization when containers and waste characteristics interfere with sterilization.

Increasing the Exposure Period Temperature

A possibly more effective way of achieving greater processing efficiency is by increasing the temperature of the exposure period. A technical publication of one manufacturer advises that steam sterilization of infectious waste is much more effective and efficient at 270°F rather than 250°F, which is commonly used at health care institutions.[4] The use of 250°F for autoclaving infectious waste is probably the result of following the parameters typically used for steam sterilizing supplies, which were adopted to avoid problems of melting. With wastes, however, thermal instability is not a serious concern and can be controlled with secondary containment. As discussed, melting of waste containers can actually facilitate steam penetration. At 270°F, waste load heating is faster and more thorough, which can improve sterilization efficiency. Before attempting higher temperatures, check with the sterilizer manufacturer as to its pressure limitations—some older models may not be designed to function at the 28–30 psi required for 270°F operation. Many modern autoclaves can, however, operate at 270°F; some even have "flash" cycles designed for higher temperatures.

Several of the above solutions are often used together to achieve sterilization and greater processing efficiency. For example, one institution found through testing that sterilization of a load of needle boxes required (1) raising the temperature from 250 to 270°F, (2) lengthening the pre-vacuum stage from 3 to 6 minutes, and (3) using a 45-minute exposure period instead of the 25-minute exposure period routinely used for bagged infectious waste. Note that under ideal conditions, only 2 minutes at 270°F is required for sterilization. Be aware that these conditions are specific for a certain type of needle container, load configuration, and sterilizer: do not assume these conditions assure sterilization for other situations.

Other Procedures That May Facilitate Sterilization

The necessity of containing infectious waste to protect waste handlers directly conflicts with the need to allow steam penetration into the waste load. This can be a serious dilemma. Additional procedures have been reported to overcome this problem, but their practical use requires considerable caution to minimize the risk of exposure of workers to infectious agents. Rutala et al. reported that waste heating was more efficient when bags were opened and their sides were folded down.[2] Similarly, EPA suggests that "bags should be opened and bottle caps and stoppers should be loosened immediately before placement in the steam sterilizer."[11] Rutala et al. also describe punching holes in the tops of closed bags before autoclaving. Other sources have advocated the addition of water to waste bags that are to be steam sterilized for the purpose of promoting internal steam generation.[3] Water can be carefully added at the point of waste generation, before the bag is closed for transport to the sterilizer, so the risk of worker exposure should generally be no more hazardous than adding waste. However, Rutala et al. question the efficacy of adding water.[2]

The above practices may possibly generate aerosols or require some handling of uncontained infectious waste, either of which can be hazardous. Any procedure that may pose a risk to waste handlers or autoclave operators should only be considered after previously described alternatives have been unsuccessful in sterilizing infectious waste.

DEVELOPMENT OF STANDARD OPERATING PROCEDURES

The development of standard operating procedures follows these steps.

Define a Standard Load

The conditions required for steam sterilization can vary considerably according to the waste and load characteristics. To overcome this variability for routine operation, a standard load for steam sterilization should be established. The load should be defined by included waste category (or varieties of wastes), excluded wastes, type of containment, load weight, and configuration. If relatively small amounts of infectious waste are steam sterilized, use a worst-case load to establish one standard operating procedure for all wastes. A worst-case load would be the largest, heaviest, and most varied and inclusive (of waste types) load that would practically be considered as one load. Worst-case loads, however, *should* employ efficient containers, load carriers, and configuration. Consider differences in waste mass, heat capacity, and ability to conduct heat.

Generators of large volumes of infectious waste need to survey their institution and categorize wastes according to the conditions required for sterilization. As a result, two or more standard loads may be designated. Hospitals generate such a varied mixture of wastes that it is difficult to define a single standard load. Instead, establish nominal conditions for a load consisting of routinely sterilized wastes, and establish more rigorous conditions for other wastes that are more difficult to steam sterilize. For example, it could be decided that needle boxes require unique sterilizing conditions and therefore should be sterilized separately. Standard loads and procedures should then be established just for needle boxes.

Test for Conditions That Achieve Sterilization

The use of retorts and autoclaves to treat infectious waste is based on the standard of sterilization. In these processes, steam is not used to disinfect or to sanitize wastes. Rather, sterilization is the goal of steam sterilization: destruction of all vegetative microorganisms and their spores. Because the identity of the infectious agents in the waste varies, or in many cases is unknown, a standard biological indicator placed in the waste load is used to establish standard operating conditions and test for sterilization. Thermocouples placed in the waste load may imply that sterilization temperatures have been attained, but they do not verify sterilization. Chemical indicators also provide evidence of sterilization, but they have not been shown to integrate time and temperature, or measure direct steam contact accurately enough to replace the use of biological indicators for the purpose of developing standard operating procedures and periodic quality assurance.

For steam, sterilization is defined as the destruction of spores of *Bacillus stearothermophilus*, which was chosen because of its resistance to steam sterilization.[12*] To test for steam sterilization, the spores of *B. stearothermophilus* are placed in the waste load before loading and removed after the process is completed. The spores are then cultured and incubated for about a week. *B. stearothermophilus*

---

*An alternative, less accepted criterion of measuring steam sterilization of infectious waste has been offered by Rutala et al. "It seems reasonable to support the selection of an autoclave processing time that provides consistent destruction of pathogenic vegetative and sporeforming bacteria but that does not necessarily eliminate the spores of *B. stearothermophilus*" (p. 1316). However, an alternative biological indicator standard has not been widely adopted.

spores are available commercially on strips or disks, or in ampoules containing culture medium and an indicator that, after incubation, changes color if there is growth because sterilization was unsuccessful.

The proper placement of biological indicators in the waste is important. Choose locations in the waste load that are the most difficult for the steam to penetrate. First consider the center of the waste load, as it is most insulated and most difficult to heat. Other good locations for biological indicators are near the bottom of the load (where air is difficult to remove), in cool areas of larger chambers (e.g., near the drain or door), and next to thermocouples (to correlate time/temperature profiles with sterilization). Simulate actual waste processing conditions as much as possible. For waste that is contained, test with a biological indicator that is contained (e.g., place the indicator in an empty needle box). One investigator inserted spore strips in empty petri dishes and into the lumens of pipets.[4] Place cultures in cavities and depressions (where air is likely to pool) or in areas from which air would be difficult to remove.

Establish Standard Conditions, Equipment, and Procedures

Waste characteristics and the size of the load all affect sterilization. Several combinations and variations should be tested. Maximum load size may differ with waste type. Experiment with various types of waste containers. Most institutions have had to try several types of bags before selecting one that meets their needs of worker protection, containment, and efficient steam sterilization. When choosing load carriers, such as trays or racks, note that metal is a better conductor of heat than plastic. Test alternative load configurations and load carriers. Try racks and load carriers that allow more space between containers.

Operational parameters most frequently varied are the pre-vacuum stage (if the autoclave is equipped with the feature) and the temperature and duration of the exposure period. If the autoclave has thermocouples that can monitor waste load temperatures, review time/temperature profiles to identify cool spots in the load (e.g., center, bottom). To allow sufficient time for sterilization, temperatures throughout the load should equilibrate at or near the chamber set point well before the end of the exposure period. During these investigations, time/temperature profiles can be very useful in determining standard operating procedures by identifying cool spots in the chamber and waste load, by determining the warmup rate of different waste containers, and by identifying the optimum load configuration.

This testing also can help establish conditions that save energy and labor costs. For example, higher temperatures may permit shorter processing cycles. Also, specifying a maximum load size that ensures sterilization will define the institution's treatment capacity and help in planning for waste management. These tests should establish optimal wastes, containers, load sizes, load configurations, and operational parameters that assure consistent sterilization as determined by the biological indicators.

Write Standard Operating Procedures

Once the conditions required for sterilization are defined by testing, procedures should be written to document proper sterilizer operation. The procedures should specify all parameters that, through testing, have been determined to ensure consistent sterilization. Include a description of the appropriate waste categories, and a standard load. Also specify containers, load carriers, the maximum load weight, details of the pre-vacuum stage, exposure period duration and temperature, and the speed of the warmup and cooldown periods for each waste type. Include photos of the waste categories and how the waste load should be configured.

The procedure should require the use of autoclave tape or another alternative chemical indicator of sterilization that changes color or markings after autoclaving. Use these with every load, preferably at several places within the load.

It is a good idea to include a facility engineer in this review to integrate the physical operational requirements. Also involve the institutional purchasing agents in this step. They need to include the proper specifications in their container purchase orders and to realize that products cannot be substituted without additional testing and approval.

STEAM STERILIZATION IN PRACTICE

Steam sterilization is the traditional method of sterilization in hospitals and other health care institutions. Autoclaves have proven to be reliable for daily sterilization of hospital supplies, equipment, and instruments. However, there are important differences between autoclaving of supplies and steam sterilization of infectious waste.

Physical Requirements

Although it is cost-effective to use an autoclave for multiple uses, the necessity of infection control dictates that steam sterilization of infectious waste be kept separate from routine autoclaving of supplies. Ideally (and if at all possible), an autoclave should be dedicated to treating infectious waste and kept separate from autoclaves used to sterilize supplies. If the autoclave used for infectious waste must be used for other purposes, special precautions should be taken when alternating uses. Work areas should be designed to keep supplies and wastes separate. A double door sterilizer may help to separate infectious waste and supply sterilization activities. Alternatively, certain shifts or personnel can be designated solely for treating infectious waste. One laboratory keeps infectious waste treatment separate from other autoclave uses by assigning staff from the daytime shift responsibility for preparing waste loads, while late-night personnel are responsible for waste loading, processing, and emptying.

It is important to allocate adequate space for waste processing. Since autoclaves are operated in a batchwise manner, separate space is needed for treated and untreated waste storage. If putrefaction of untreated waste is a potential problem, ample refrigerated storage space should be made available, preferably with temperatures below freezing.

Many microbiology laboratories have autoclaves for infectious waste sterilization located in or adjacent to the laboratory. This location makes sense for a sterilizer dedicated to treating high-risk waste from a single source. For treating infectious waste from multiple sources, such as in a hospital, the optimal location will likely be near the loading dock where sterilized waste leaves the building. Loading dock areas tend to be crowded, however, and it is prudent in all cases to keep infectious waste management operations isolated from other activities as much as possible.

An important advantage of steam sterilization for treating infectious waste is that, unlike incineration, local and state licenses or permits are rarely required. The installation of a steam sterilizer requires no political siting considerations.

Labor Requirements

Because autoclaves operate in a batchwise manner, labor requirements for steam sterilization are typically sporadic. Labor is needed before the processing cycle to assemble the load, then to operate the load, and after sterilization to disassemble the load and dispose of the sterilized waste. With modern autoclaves that have automatic or semiautomatic operation, the employee need only start the cycle and can then be available for other work during the pre-vacuum, steam charging, sterilization, and steam discharging stages. As a result, autoclave operation is an employee's secondary job in many cases. In busy hospitals, such an employee will also dispose of sterilized waste, keep records, and prepare the next load.

Operational schedules can vary from single-shift to more efficient multiple-shift operation. Several-shift operation may allow the purchase of a smaller, less expensive sterilizer.

Operator Training

Modern autoclaves with computerized controls and programmable operation make training easier, but successful sterilizer operation still requires operator training and annual refresher training. Learning and understanding the written procedures should be the focus of operator training. Operators and their supervisors need to understand the principles of steam sterilization, interfering factors, and the practices incorporated into the standard operating procedures to overcome them. Explain the proper use of chemical indicators and their limitations.

Operators should be instructed as to the importance of proceeding slowly through the steam charging and discharge stages. If steam is charged into the chamber too quickly, infectious agents can aerosolize and exit the drain of the autoclave during warmup. Likewise, a hurried steam discharge stage risks release of infectious agents via the vent. As the principal means of preventing releases, filters should be present to trap aerosols and particulates during the charging, venting, and pre-vacuum stages, and many sterilizers control these stages automatically.

Inexperienced operators sometimes place undue importance on chamber pressure. Remind operators that steam sterilization requires attainment of a specified temperature. Pressure is incidental and can be misleading when the air is not completely displaced by steam. The pressures typically employed for steam sterilization can be achieved by a mixture of steam and air, which is the result of incomplete displacement of air by steam. For example, Figure 6.2 shows two loads at the same pressure (20 psi), but with different temperatures.

Recordkeeping

Maintain records of biological indicator tests, chemical indicator readings, maintenance, filter changes, and inspections. A logbook should be kept to record each cycle, load description, operator, day, and time. Keep temperature charts and other operational charts with other records and store them for at least three years (or as necessary for review by regulatory authorities or other inspections).

High-Risk Waste

Special procedures are prudent when sterilizing high-risk waste or wastes that have proven difficult to sterilize. Appropriately stringent conditions include prolonged pre-vacuum stage and exposure period, special handling (e.g., only by the biosafety officer), use of biological monitors for each load, and holding the sterilized waste until test results show sterilization. Decontaminate external sterilizer surfaces after each shift under routine conditions and after each load with high-risk waste. In some cases, high-risk autoclaved wastes are disposed of by incineration.

DISPOSAL OF STEAM STERILIZED WASTE

The difficulty of waste disposal is the most serious disadvantage of steam sterilization as an infectious waste treatment method. Steam sterilization often results in little or no visible changes to the waste. After autoclaving, infectious waste still looks like infectious waste. It may be hard to convince solid waste haulers that the waste has been sterilized and it is safe to handle. This concern varies greatly by state and locale. Medical institutions in large urban areas seem to have the most difficult problems with the acceptance of steam sterilized waste as normal trash. For some institutions, this problem is serious enough to disqualify steam sterilization from consideration for routine management of most infectious waste. In other areas, waste haulers handle autoclaved waste with no objection, but this is becoming less common.

Trust is an important factor. To accept the treated waste, solid waste haulers and facilities need to be confident that the institution has an effective steam sterilization program. One institution overcame this problem by giving their municipal waste handlers a tour of their facility and operations, explaining how standard operating procedures were developed and how they are enforced within the institution through its quality assurance program.

Disposal of sterilized needles and syringes is a special problem. Not only do sterile needles usually look like contaminated needles, but even sterile needles and syringes can cause serious puncture wounds to waste handlers. Also, needles are among the most offensive of wastes in the public eye. Although steam sterilization is an excellent method for treating needles contaminated with infectious agents, autoclaved needles need to be encapsulated or destroyed prior to disposal. At a minimum, all needles must be disposed of in puncture-proof containers that have been securely closed. But even this can be unsatisfactory if compacting of the needle container at the loading dock or in a truck prior to landfilling results in a spill. Some institutions double-box their waste for this reason.

To distinguish treated waste from untreated waste, some locales require the marking or defacing of waste labels and biohazard symbols on the containers. Chemical indicators attached to the waste load, such as tape or labels that change appearance, can only be trusted by those few employees who know how to interpret them. For many institutions, defacing warning labels or marking treated waste as "sterilized" or "autoclaved" is a useful way of distinguishing treated and untreated waste.

To address the problem of distinguishing treated waste from untreated waste, the EPA regulations (currently applicable in only a few states: see Appendix A) require that infectious waste be destroyed in order to be exempt from regulation. Destruction, according to EPA, means "waste that has been ruined, torn apart, or mutilated through processes such as thermal treatment, melting, shredding, grinding, tearing or breaking, so that it is no longer generally recognizable as medical waste. It does not mean compaction."[13] Although grinding waste after steam sterilization has been a relatively uncommon procedure, some facilities routinely incinerate infectious waste after autoclaving it to overcome these aesthetic concerns.

QUALITY ASSURANCE AND QUALITY CONTROL (QA/QC)

Critical and complex procedures, such as steam sterilization of infectious waste, merit special attention to quality control. A quality assurance program is not only a good idea but may be required for institutional accreditation.

The first job in quality assurance is the responsibility of the autoclave operator. After every cycle, he or she should check the temperature plots prior to opening the sterilizer door to make sure that the exposure period temperature was achieved and that there were no anomalies during the cycle. Autoclaved wastes must not be disposed of unless the chemical indicators indicate successful sterilization. All sterilization failures should be reported to the biosafety officer or facility engineer.

Oversight of the quality assurance program should be the responsibility of the biosafety or infection control officer, whose duties should include oversight of waste surveys and testing to develop the standard operating procedures, writing (or approving) the standard operating procedures, and assisting with operator training.

Review of Equipment and Instrumentation

Timers and temperature monitors need to be checked and calibrated on a regular basis. Periodic maintenance by the manufacturer is also prudent—no one knows more about what problems to look for. Other routine maintenance includes replacing filters, gaskets, and thermocouple connections, and cleaning drains. Running an empty chamber through a standard cycle is a good periodic measure of consistent operation. Temperatures should be uniform in all areas for any empty cycle. From year to year, the time/temperature profiles and vacuum and steam pressure readings should not vary when the chamber is empty.

Review of Standard Operating Procedures

Standard operating procedures also require periodic review. Procedures that ensure consistent sterilization are likely to vary as waste types change (e.g., when a new type of waste needle collection box is used), when new thermocouples and gauges are installed, etc. Such occurrences should prompt retesting of sterilization conditions. Some states require hospitals to regularly test the effectiveness of steam sterilization using spore cultures (e.g., once a month and when procedures change).

Supervisors have the critical responsibility of daily oversight of employees and their practices. However, spot inspections of sterilizer operations should be done by the biosafety officer to ensure that the standard operating procedures are being followed. Chemical indicator results and other records should be reviewed. Audit waste disposal practices. Finally, document all quality assurance activities.

REFERENCES

1. Perkins, J. J. *Principles and Methods of Sterilization in Health Sciences,* 2nd ed. (Springfield, IL: Charles C. Thomas, 1983).
2. Rutala, W. A., M. M. Stiegel, and F. A. Sarubbi, Jr. "Decontamination of Laboratory Microbiological Waste by Steam Sterilization." *Appl. Environ. Microbiol.* 43(6):1311–1416 (1982).
3. Lauer, J. L., D. R. Battles, and D. Vesley. "Decontaminating Infectious Waste by Autoclaving." *Appl. Environ. Microbiol.* 44(3):690–694 (1982).

4. Cooney, T. E. "Techniques for Steam Sterilizing Laboratory Waste," AMSCO Waste Processing Technical Report (1988).
5. *State Infectious Waste Regulatory Programs* (Lexington, KY: The Council of State Governments, 1988).
6. "EPA Guide for Infectious Waste Management," U.S. Environmental Protection Agency Office of Solid Waste and Emergency Response, National Technical Information Service, PB 86-199130 (May 1986), p. 5-2.
7. *Biosafety in the Laboratory: Prudent Practices for the Handling and Disposal of Infectious Materials* (Washington, DC: National Academy Press, National Research Council, 1989), p. 39.
8. Everall, P. H., and C. A. Morris. "Failure to Sterilize in Plastic Bags," *J. Clin. Pathol.* 29(12):1132 (1976).
9. Dole, M. "Warning on Autoclavable Bags," *Am. Soc. Microbiol. News* 44(6):283 (1978).
10. "EPA Guide for Infectious Waste Management," pp. xiv and 5-1 through 5-5.
11. "EPA Guide for Infectious Waste Management," p. 4-5.
12. *The United States Pharmacopeia* (Rockville, MD: United States Pharmacopeial Convention, Inc., 1989), pp. 1625–6 and 1705–7.
13. *Code of Federal Regulations,* Title 40, Part 259.10.

CHAPTER 7

Incineration of Infectious Waste

This chapter is not meant to be a complete reference for incineration engineering, but rather a guide to the issues and aspects to consider for infectious and medical waste incineration. See references 1, 2 and 3 for detailed discussions of incineration engineering.

INCINERATION: *THE* SOLUTION?

In many ways, incineration appears to be the ideal solution to waste management. An incinerator has the capability of destroying hazardous components of a waste, of reducing a waste's volume by leaving only ash to dispose of, and of recovering energy from the waste—all at the same time. In fact, incineration is firmly entrenched as the preferred method of treatment for many types of infectious wastes.

Many attorneys who specialize in environmental law favor incineration too. For the price, few other processes have the destruction capabilities of incineration, and destruction is the shortest path to reduced liability. Incineration also does the best job of all treatment methods of destroying the evidence of infectious waste generation. If incineration is done properly, only ash remains; labels and patient bracelets don't exist to trace the waste to its generator. Ash itself is a visual confirmation that treatment has taken place. And ash that contains no recognizable residuals is a good indicator that incineration was conducted properly, that the waste was exposed to conditions sufficient to destroy infectious agents.

There is a rising sentiment, however, that waste incineration has serious drawbacks.[4] Critics argue that incineration merely transfers pollutants from solid media (waste) to the air (via stack emissions), water (via scrubber discharges), and ash. While waste burns, chemical transformations take place. Organic compounds recombine, sometimes into chemicals that are toxic at extremely low concentrations. Polychlorinated dibenzo-*p*-dioxins (referred to as dioxin) and polychlorinated dibenzofurans (or furan) are two such toxic combustion products and, according to the Congressional Office of Technology Assessment (OTA), "concentrations of both dioxin and furans are considerably higher in hospital incinerator fly ash than in municipal incinerator fly ash."[5]

Trace amounts of toxic metals in the waste may concentrate in the ash, sometimes creating ash that is also a hazardous waste. With the exception of mercury (which stays in the vapor), metals volatilize, condense on particulates, and concentrate in the fly ash. Since hazardous waste cannot be disposed of in a municipal waste (sanitary) landfill, significant additional expenses may be incurred for ash disposal. Because of these concerns about ash disposal, OTA concludes, "it is unlikely that disposal of incinerator ash in existing municipal landfills will continue to be allowed."

Incineration is also used as an excuse, some say, to avoid the most preferable method of waste management: waste minimization at the point of generation. Gaseous hydrogen chloride and nitrogen oxides, alleged to contribute to acid rain, are also released during incineration. Carbon dioxide, a product of combustion of fossil fuels (i.e., gas, oil, coal) as well as wastes, may result in global warming, some fear. Surely, the contribution of these pollutants to the environment from incineration of waste, and especially medical waste, is minuscule compared to emissions from burning fossil fuels. (Dioxin and furan emissions from waste incinerators are a possible exception.) However, the Office of Technology Assessment speculates that infectious and medical waste incinerators may have a greater impact on public health because they are commonly in closer proximity to residential areas than those larger sources.

Accordingly, it is difficult to find a site for a new incinerator. People living nearby fear that their health may be affected and raise the above arguments against incineration. If the community perceives that the incinerator causes an adverse impact, the value of adjacent property may be lowered. Thus, many communities have loudly voiced their opposition to incinerator construction, a syndrome given the acronym of NIMBY (Not In My Back Yard).

The authors contend that incineration remains a reliable and necessary technology for managing medical wastes. Incineration is one of the few rational ways to treat some wastes, such as body parts, tissue from pathology laboratories, and animal carcasses. With proper engineering, source separation, thorough operator training, close supervision, air pollution control devices, and special ash disposal procedures, an incinerator's environmental impact can be insignificant. It is likely, however, that the cost of such careful incineration practices will be significantly more than many institutions are experiencing today.

HOW INFECTIOUS WASTE INCINERATORS WORK

Incinerators for infectious waste are supplied under various names: medical, infectious, or pathological waste incinerators. 90% of those installed during the last two decades are properly called controlled air incinerators.[3]

Controlled Air Incinerators

The physical components of a controlled air incinerator include, in process order, a primary combustion chamber (into which waste is fed), a secondary combustion chamber (to incinerate gases that have volatilized from the primary chamber), and a stack to vent combustion gases. If an air pollution control system is used to remove particulates and acid gases, scrubbers are installed after the secondary combustion chamber. To recover heat, a boiler may be added after the secondary chamber and before the scrubber. Options to facilitate throughput include automated systems for waste feed (e.g., a hydraulic ram) and ash removal. See Figure 7.1 and Table 7.1.

A controlled air incinerator normally operates with restricted air flow into the primary chamber (i.e., starved air). The combustion temperature of a controlled air incinerator is controlled by manipulating the air supply; more air increases

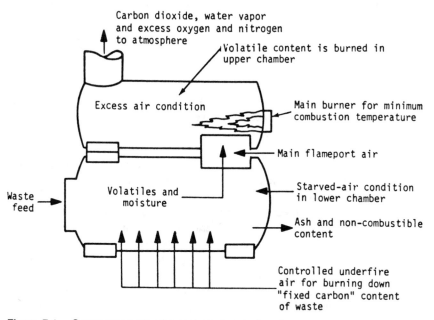

Figure 7.1. Component parts of a controlled air incinerator.

Table 7.1. Components of an Infectious Waste Incinerator[a]

Component	Purpose	How It Works
Primary combustion chamber	Dry, heat and convert waste to gases; some oxidation and pyrolysis may occur.	Auxiliary fuel (e.g., natural gas) is used to raise the chamber temperature to 1600–1800°F to initiate and maintain combustion.
Secondary combustion chamber	Oxidize gaseous waste into carbon dioxide and water.	With sufficient oxygen present (supplied as air), 1800°F and above efficiently converts organic components of the waste into carbon dioxide and water.
Boiler[b]	Recover heat from exhaust gases.	A heat exchanger converts water to steam; steam can be used for heating and cooling.
Air pollution control devices[c]	Remove hydrogen chloride, other acid gases, and particulates.	See Table 7.2.
Stack	Disperse exhaust gases.	Releases exhaust gases from the secondary combustion chamber at a height and location that allows dilution to safe concentrations.

[a]Typical for a state-of-the-art controlled air incinerator. Listed in order, from waste input to gas emissions.
[b]Optional; may need to cool exhaust gases for scrubbing if not used.
[c]May not be present on older units, but usually required for new installations.

the operating temperature. Due to the lack of oxygen, the primary chamber of a controlled air incinerator functions to dry, heat, and volatilize the waste. The primary chamber begins the combustion process, while the secondary chamber completes it.

Chemical waste incinerators permitted by EPA under the authority of the Resource Conservation and Recovery Act (RCRA hazardous waste incinerators) are typically either strictly liquid injection units (for waste solvents) or of rotary kiln design (which can directly treat large solid objects—entire drums can be fed into the incinerator). Some RCRA incinerators, including the rotary kiln types, also accept infectious wastes. Rotary kiln units are being considered as an alternative to controlled air units for larger new installations. There has been some success in using a small rotary kiln incinerator for infectious waste.[6]

Although some municipal waste incinerators can efficiently treat infectious wastes, most facilities lack the special handling procedures necessary for infectious waste and therefore require that such waste be sterilized before it will be accepted.

Incineration Chemistry

An incinerator is really a large-scale chemical reactor. Incineration represents thermal oxidation, a type of chemical reaction that takes place at high temperatures, after the waste has been converted to gas. The process is the same as combustion of any fuel, as in a car or home furnace. Medical waste may have sufficient fuel value to serve as the fuel in an incinerator, although auxiliary fuels (gas or oil) are used to heat the waste and aid combustion.

In an incinerator, organic (carbon-based) chemicals in the waste are volatilized, oxidized, and released to the atmosphere as carbon dioxide and water. Minerals, metals, and uncombusted organic materials are left as ash, which weighs 10–30% (depending on the type of waste) of the weight of the original solid waste. Sources of organic material in the waste include infectious agents, animal tissue, blood, other biochemicals, paper, and plastics. Like all chemical reactions, optimal conditions are required for efficient incineration. Incineration requires five conditions:

- oxygen in sufficient amount to completely react with all the waste (supplied as air via blower fans)
- atomization and volatilization of the waste into very small particles and gases. Reactions can only take place in the gas phase or on surfaces that allow oxygen to directly contact the waste.
- turbulence and mixing in the incinerator (to bring the atomized waste in contact with the oxygen)
- temperature sufficient to volatilize the waste, support the self-sustaining condition of combustion, and allow reactions to occur in a reasonably short time
- sufficient time under these conditions to allow the reaction to take place. Retention time is the length of time the gaseous waste spends in the combustion chambers, not including the time required for volatilization of the waste.

Many people make the mistake of considering only the time and temperature requirements for incineration without due consideration of the other three factors. For example, combustion with insufficient oxygen results in partial oxidation and pyrolysis. Incinerators operated at commonly quoted standards for time and temperature have been found to produce ash with easily recognizable (untreated) waste. At the molecular level, chemicals formed under less-than-optimal conditions are called products of incomplete combustion (PICs). They include highly toxic dioxins and furans (mentioned previously), polyaromatic hydrocarbons (PAHs), and other polyaromatic organic matter (POMs). For example, while controlled air incineration minimizes particulate emissions, sub-stoichiometric air (less oxygen than necessary for complete combustion) can foster dioxin formation. All of these chemicals are increasingly the target of pollution control regulations.

Destruction Efficiency

The critical measurement of incinerator effectiveness is destruction efficiency. Destruction efficiency is measured in several ways, depending on the hazardous material on which it is based. The destruction efficiency for a chemical waste incinerator is measured as the percentage of a difficult-to-destroy chemical in the waste that is not emitted from the stack. For infectious waste, destruction efficiency is measured by the ability to destroy infectious agents, resulting in no infectious agents remaining in the ash or being emitted from the stack.

Although testing for destruction efficiency is a standard requirement for chemical waste incinerators, only a handful of medical waste incinerators have been thoroughly tested to assure that they destroy infectious agents. Citations in EPA's "Draft Manual for Infectious Waste Management"[6] show, however, that infectious agents can remain in the ash or be released from the stack of medical waste incinerators.[7-10]

Infectious waste incinerator tests are done using the spores of *Bacillus subtilis* var. *niger* (*globigii*), which are especially resistant to heat.[11] The biological indicators are added to the waste, and the ash and stack emissions are monitored for their presence. The State of Pennsylvania has developed a standard test for incinerator ash sterility.[12] Other possible sources of release may merit testing as well, such as scrubber wastewater.

High destruction efficiency is achieved by exposing the waste to optimal incineration conditions, as described in the previous section. Incinerators are designed to provide excess oxygen (only in the secondary chamber for controlled air units), atomization, and turbulence.

Optimal conditions also depend on waste characteristics. Large amounts require more oxygen. Wet waste will take longer to heat and volatilize. When comparing a waste that occupies the same volume, waste with greater surface area and having a lower density will heat more rapidly and thoroughly. For example, a large animal carcass is massive and a poor conductor of heat and has a small amount of surface area in relation to its volume. The same can be true for compacted waste—the density is high and effective surface area is lowered, thereby making heat conduction less efficient. Incineration is difficult for animal tissue, (especially carcasses); long burndown periods are necessary to thoroughly heat and combust the waste.

Longer retention times or higher temperatures don't necessarily increase the destruction efficiency. Because retention time depends on the combustion gas flow rate, and flow rate depends on temperature, retention time and temperature have an inverse relationship; increasing temperature usually decreases retention time. Sometimes a lower temperature will result in a better destruction efficiency by allowing a longer retention time. Because of the great variation in institutional waste streams and incineration engineering, conditions necessary for destruction can vary with each institution and sometimes even with each load.

Still, time and temperature are important conditions; they are often the basis for some state regulations. Reference time and temperature conditions for infectious waste incinerators have increased over the years as shown in Table 7.2.[12]

Table 7.2.   Standard Reference Combustion Temperatures for Infectious Waste Incinerators[12]

Combustion Chamber	Previous	Newer
Primary	1400–1600°F	1600–1800°F
Secondary	1400–1600°F	1800+ °F
Secondary Chamber Retention Time	1/4 to 1/2 seconds	1 to 2 seconds

Caution: Destruction requires additional conditions; see text.

Burnout and Burndown

Another indication of incineration efficiency is burnout. Burnout measures the completeness of combustion in the bottom ash (ash that settles on the floor of the incinerator). Ash that contains significant amounts of uncombusted waste is a sign of poor burnout; in such cases gross inspection of the ash may reveal large pieces of partially burned, recognizable material. Two standard measurement methods exist as published by ASTM and the American Society of Mechanical Engineers (ASME).

For most medical waste incinerators, a significant amount of waste oxidation occurs on the surface of the waste mass. Bulk waste does not have the large available surface area of atomized waste. The rate of surface oxidation, atomization and volatilization is by far the slowest step in combustion of the waste, and thereby determines the rate at which waste can be fed into the incinerator. Some incinerators have special features to facilitate atomization and volatilization, such as shakers and agitators. Some ash transfer and removal systems may also serve this purpose.

Poor burnout is a common observation at medical waste incinerators.[13] Causes of insufficient burnout include poor incinerator design, certain waste characteristics (e.g., high density, low surface area), excessive waste feed rate, insufficient burndown time (time between batches of waste), and premature shutdown (before complete combustion of the waste). For the last three items, poor standard operating procedures or operator training are the overall causes.

REGULATION OF MEDICAL WASTE INCINERATORS

Infectious waste incinerators are regulated under various local, state, and federal laws. Federal regulations are limited to standards for particulate emissions for very large incinerators and the recordkeeping and reporting provisions of EPA's medical waste tracking rules, which apply to a few states as of 1989. (See Chapter 2.)

State Air Pollution Regulations

Air quality rules are enforced by the states in most cases, and the states have taken the lead in regulating infectious waste incinerators.

Although opacity guidelines have been issued by some states, most states did not have opacity and particulate emission limits for hospital incinerators as of 1987.[13] Still, many states have or are developing rules that are more stringent than the federal standard in that they cover smaller air pollution sources and more pollutants (e.g., chlorine, metals, volatile organic chemicals), and the emission limits are lower. In some states it is not practical or allowed to operate an infectious waste incinerator without the inclusion of air pollution control equipment that removes hydrogen chloride and particulate matter.

New state incinerator regulations have focused on the emission of products of incomplete combustion, and waste incinerators have been shown to be significant sources. Emission limit regulations are either performance-based or engineering-based. Traditional performance-based rulemaking specified emission limits for pollutants. Increasingly, emission standards are specifying that incinerators employ the best available control technology (BACT)—in some cases, regardless of cost. Thus, as pollution control technologies improve, incinerator emissions are commensurately restricted. In this way, air quality regulations continuously reflect state-of-the-art pollution control capabilities. Such an environmental policy replaces risk assessment as a means of establishing acceptable releases with the goal of preventing impacts to the greatest extent possible.

Some new state rules conditionally exempt smaller incinerators or incinerators already existing, or delay their compliance date. This is called a ''grandfather'' provision. Check with the state air quality agency to determine what standards apply.

Medical Waste Tracking Rules

EPA's rules for medical waste require keeping records of incinerator operation (such as an operator's log of throughput) and periodic reporting to EPA of the volume and types of medical waste incinerated. See 40 CFR 259.60–2 (Appendix A).

DECISIONS: RETROFIT, REPLACE OR BUILD

In many cases, new federal, state, and local rules for infectious waste management and incineration will require a reevaluation of a currently used incinerator or will result in a proposal to construct a new one. There are some indications that controlled air incinerators, popular for their ability to meet the particulate standards of the '70s and '80s, may not be successful in meeting the more stringent standards of the '90s.[14,15] Take advantage of exemptions for older and smaller incinerators, if applicable, but plan for improvements to meet the new standards, replacement of the existing unit, upgrade of air pollution control equipment, or offsite shipment.

Evaluating an Existing Incinerator

Many hospital have incinerators that were installed in the '60s and early '70s. Today, these units fall far below the standards for efficient combustion. It doesn't always take an expert to spot evidence of such a problem: low operating temperature, partially burned waste remaining in the ash, high stack opacity, or frequent upsets. Be suspicious of an older incinerator, or one that lacks a secondary combustion chamber to ensure destruction of infectious agents.

If space is available and the engineering is feasible, it may be possible to retrofit an older incinerator for more efficient operation and air pollution control. Consult an incinerator expert and institutions that have retrofitted an incinerator as to the feasibility of retrofitting. The rule of thumb is that retrofitting may be advisable for incinerators less than five years old if the cost of the improvement is less than 60% of the price of a new incinerator.[12]

In many cases, modification of an existing incinerator is not a cost-effective way of meeting new incinerator emission limits. This is especially true with incinerators more than 20 years old. There are three other options: offsite treatment (discussed in Chapter 16), alternative treatment technologies (see Chapters 5, 6, and 8), and construction of a new incinerator.

Building a New Incinerator

An incinerator and related equipment is a significant capital outlay, and the purchase should not be made solely on the advice of a manufacturer. Regulations for licensing a new incinerator may require stack tests, demonstrating a need for the facility, public hearings, and evaluating the environmental impact of incinerator operation. Incinerator planning and design decisions are treacherous. Over the last two decades, new incinerators have been of controlled air design. But as noted by L. G. Doucet, "despite its increasing importance and advancements, controlled air incineration technology has a somewhat questionable track record. Hundreds of installations perform poorly or not at all . . . roughly 25 percent of all new systems do not operate properly.''[3] In addition, as mentioned in the following section, a controlled air incinerator may have difficulty meeting newer air pollution control standards.

The authors strongly recommend that a consultant or engineering consulting firm with extensive experience in incinerator engineering be retained to assess institutional needs, draft the bid specifications, and advise on the feasibility and specifications of such a facility, including equipment, design, testing protocol, and operation. Also seek out other institutions that have recently installed an incinerator and solicit their advice.

Some incinerator consultants advise that new hospital incinerators should meet EPA's 99.99% destruction and removal efficiency (DRE) standard for chemical waste. They reason that infectious wastes may fall under the same standards some day, and the incinerator could be permitted to burn chemical wastes in addition to medical waste. This may prescribe a secondary combustion chamber temperature of 2200°F. High temperatures and long retention times for the secondary combustion chamber require an unusually large unit. Also, combustion at these temperatures produces nitrogen oxides, a pollutant of concern. Required RCRA instrumentation is extensive: carbon monoxide monitoring (to detect combustion upsets); gauges to measure waste feed rate, auxiliary fuel, and combustion air; and automatic shutdown controls. Other engineering differences between infectious and chemical waste incinerators are necessary because of differences in waste characteristics (e.g., Btu value, physical state).[16]

The authors believe it is advisable to purchase an incinerator that exceeds current standards for burning infectious waste. However, the costs to meet RCRA standards are likely to be excessive (few medical institutions will be able to afford it) and do not appear to be justified at this time.

Cost of Construction and Operation

Manufacturing costs, engineering constraints, and a lack of enough buyers make small incinerators disproportionately expensive when compared to larger incinerators. The smallest controlled air incinerator available today is rated at about 1000 lb/hr.

If air pollution control devices must be included to meet existing or planned emission standards, it may be cost-effective to also add a heat recovery boiler. In some states, energy credits are available to help pay for the cost of the heat recovery system. The boiler can serve as a means of reducing the exhaust gas temperature. (Most air pollution control devices require or work best at lower temperatures.)

Even when additional air pollution controls are not required, heat recovery is a worthwhile option to consider. This first requires that the steam produced by the boiler be used efficiently. If there is sufficient demand for steam, heat recovery also opens the door to reducing incinerator operating costs by incineration of normal solid waste in addition to the medical wastes. This larger waste volume may bring the incinerator size into the most favorable price/capacity range.[2,17] Note that larger controlled air units may have difficulty meeting new particulate standards, while the cost of a rotary kiln incinerator becomes competitive for larger facilities.

Although all incinerators require downtime for maintenance (and ash removal for manual systems), continuous duty (operation 70–80% of the time) is most cost-effective. Alternatively, extended burning periods (e.g., overnight, several consecutive days) are most efficient and reduce refractory wear. Thus, larger facilities have the advantage of several economies of scale. Large facilities that require large waste volumes argue for a regional facility that serves several waste generators. This option is discussed further in Chapter 17. Alternatively, refrigerated storage facilities that allow for periodic, extended runs are likely to pay off in lower operation costs.

The cost to control environmental impacts and properly operate an incinerator is greatly underestimated (and sometimes ignored) in most proposals for incinerator construction. The costs of compliance procedures, staffing, training, and documentation are significant, and sometimes overwhelming. A large facility may need to hire an incinerator engineer to oversee operation and troubleshoot problems. At a minimum, the institutional engineer should be responsible for oversight and inspection of the incinerator's operation.

INCINERATION ENGINEERING AND OPERATION

Conditions that elicit the treatment unit's optimum efficiency define the standard operating procedures. These conditions may be comprised of a set of temperatures, pressures, and other physical measurements, and may be specific for certain waste types and other parameters.

Appropriate Waste Types

Both solid and pathological waste incinerators have difficulty with certain wastes common to medical facilities. In an incinerator, glass melts and slags, which makes ash removal difficult and can foul an automatic ash removal system or plug the underfire air ports.

Liquids pool and sometimes drain from lower openings in incinerators not specially designed to handle them. Blood and other body fluids are best handled in very small amounts or as combustible materials (e.g., paper, cloth with blood or liquids absorbed on them).

Approximately 20% of infectious and medical waste consists of plastics.[13] Many plastics contain chlorine so that, when waste containing plastics is incinerated, hydrogen chloride gas is emitted. It may be possible to control hydrogen chloride emissions (limited by law in many states) by avoiding purchase of chlorine-containing plastics. Alternatively, scrubbers to remove acid gases may be necessary. A few plastics have been found to contain toxic metals and are thought to contribute to the concentration of metals in the ash.[5]

It *is not* appropriate to dispose of chemical wastes in a pathological or infectious waste incinerator. If the waste is regulated as a hazardous chemical waste, it is illegal to burn it except in a RCRA-permitted hazardous waste incinerator. As explained in Chapter 11, small quantity generators of chemical hazardous waste are required to use a treatment facility specifically approved for such waste.

Today's state-of-the-art medical waste incinerators may be able to achieve 98% chemical destruction efficiency, but this is still below EPA's 99.99% DRE criterion for chemical waste incineration. Tests of older units have shown that chemical destruction efficiency may be as low as 85–90%.[12] See Chapter 11 for a discussion of the possibility of incinerating unregulated wastes containing organic chemicals in small amounts and low concentrations.

Instrumentation and Automation

Requirements for instrumentation and automation on medical waste incinerators vary greatly from state to state. Most states have minimal standards, such as recording operating temperature and waste feed rate.

It is prudent to monitor key combustion parameters: temperature, air supply (monitored via blowers), and auxiliary fuel feed. Records of operation can be kept by electronically connecting the monitor to a chart recorder (standard design for recording temperature) or by keeping a log book completed by the incinerator operator (standard procedure for recording waste feed rate). Some states require continuous monitors with chart recorders for stack gas opacity and, for larger units, carbon monoxide; both of these parameters signal combustion upset and inefficient destruction. Boilers and air pollution control devices both will require additional instrumentation. Newer, sophisticated systems are designed to automatically prevent waste from being fed into the unit when monitors indicate combustion upset, a breakdown, scrubber failure, or deviation from the limits of efficient combustion conditions.

Air Pollution Control Devices

Nearly all medical waste incinerators will probably require air pollution control devices (also known as scrubbers) someday. Medical waste incinerators have been identified as a significant source of pollutants,[13] and there has been a steady trend of lowering emission standards for the past two decades. Requirements are already in place in some states, while the compliance deadline for scrubber installation on other institutional incinerators will depend on federal and state regulatory activities and the extent of exemptions for small and existing units.

Several types of air pollution control devices are available to reduce particulate and hydrogen chloride gas emissions. Popular controls and their operation are described in Table 7.3. It is uncertain, however, if these controls will successfully meet the new standards. As of November 1988, J. L. Tessitore and F. L. Cross observed, "no existing single air pollution control concept has demonstrated the [new] combined limits for hospital-type facilities."[15] Further, they found that the success of dry scrubbers in conjunction with a bag house for municipal trash incinerators had not been replicated with smaller controlled air incinerators. Dry scrubbers appear to have the advantage that they are less likely to produce chemically hazardous ash due to immobilization of the heavy metals with lime.

Stack tests are sometimes required by state air quality agencies to determine if particulate emissions are below regulatory limits. Other pollutants of concern for medical waste incineration include gaseous hydrogen, nitrous oxide (above 2000°F), and products of incomplete combustion. As mentioned previously, tests to prove destruction of infectious agents are prudent. New York State estimates that required annual stack tests for medical waste incinerators will cost $5000.[18]

Table 7.3. Air Pollution Control Devices (Scrubbers) Used for Incineration[a]

Device	Purpose	How It Works
Venturi	Mixing to facilitate neutralization of acidic gases.	Neutralizing chemicals mix with stack gases.
Packed tower[b]	Remove and neutralize acid gases.	A column filled with small beads that provide a large surface area for gases and neutralizing agent to mix.
Lime injector[c]	Remove and neutralize acid gases.	Atomized lime slurry is mixed with stack gases.
Bag house[c]	Remove particulates.	Stack gases pass through filters.
Electrostatic precipitator	Remove particulates.	Charged particles are attracted to plates and removed from the stack gas.

[a]A good discussion of the many air pollution control devices in use can be found in "Air Pollution Control at Resource Recovery Facilities," State of California Air Resources Board, May 24, 1984.
[b]Component of a wet scrubber.
[c]Component of a dry scrubber.

Operation and Operator Training

Extended operational runs not only reduce refractory wear but may lower dioxin and furan emissions.[5] Refractories also require slow (several hours) and gradual warmups and cooldowns. To prevent the formation of products of incomplete combustion, never charge waste into the incinerator until the combustion chambers have reached operating temperature. For the same reason, complete waste combustion and burndown before initiating cooldown. To ensure this, some manufacturers offer automatic lock-out and shut-down options.[11]

These operating conditions all rely on the training of the operator. The operator must know the principles of combustion, key incineration parameters, proper performance of incinerator equipment and instrumentation, signs of improper operation, troubleshooting, and occupational safety. Recognizing this as a crucial component of correct medical waste incineration, EPA has issued an operator training manual.[19]

Location

The environmental impact of an incinerator depends very much on the characteristics of the site on which it is installed. Urban and residential installations often are short on space; poor stack gas buoyancy (subject to atmospheric conditions) may result in downwash into building air intakes. Dispersion modeling can be used to estimate pollutant concentrations at various points of human occupancy. Results may suggest a high stack or an alternate location.

COMPARING INCINERATION WITH OTHER TREATMENT METHODS FOR INFECTIOUS WASTE

Perhaps the most important decision in planning for infectious and medical waste management is the choice of treatment and disposal methods. Treatment is likely to be the costliest step in the waste management pathway, but more importantly, it is a step taken with great environmental and occupational risk. In Chapter 5, risks are evaluated according to criteria discussed in Chapter 1.

REFERENCES

1. Cross, F. L. and H. E. Hesketh. *Controlled Air Incineration* (Lancaster, PA: Techtonic Publishing Co, 1985).
2. Brunner, C. R. *Incineration Systems: Selection and Design* (New York: Van Nostrand Reinhold, 1984).
3. Doucet, L. "State-of-the-Art Hospital and Institutional Waste Incineration: Selection, Procurement and Operations." Paper presented at the 75th Annual Meeting of the Association of Physical Plant Administrators of Universities and Colleges, Washington, DC (July 24, 1988).
4. Environmental Defense Fund. *To Burn or Not To Burn: The Economic Advantages of Recycling over Garbage Incineration for New York City* (New York: Environmental Defense Fund, August 1985).

5. U.S. Congress. Office of Technology Assessment. "Issues in Medical Waste Management—Background Paper." OTA-BP-O-49 (Washington, DC: U.S. Government Printing Office, October 1988).
6. U.S. Environmental Protection Agency. "Draft Manual for Infectious Waste Management." EPA Publication No. SW-957. U.S. EPA, Washington, DC (September 1982).
7. Barbeito, M. S. and G. G. Gremillion. "Microbiological Safety Evaluation of an Industrial Refuse Incinerator." *Appl. Microbiol.* 16(2):291-295 (1968).
8. Barbeito, M. S. and M. Shapiro. "Microbiological Safety Evaluation of a Solid and Liquid Pathological Incinerator." *J. Med. Primatol.* 6(5):264–273 (1977).
9. Peterson, M. L. and F. J. Stutzenberger. "Microbiological Evaluation of Incinerator Operations." *Appl. Microbiol.* 18(1):8–13 (1969).
10. Barbeito, M. S., L. A. Taylor, and R. W. Seiders. "Microbiological Evaluation of a Large-Volume Air Incinerator." *Appl. Microbiol.* 16(3):490-495 (1968).
11. U.S. Environmental Protection Agency, Office of Solid Waste. "EPA Guide for Infectious Waste Management," EPA/530-SW-86-014 (May 1986).
12. Cross, F. "Evaluation of Off-Site Regional Incinerator Facilities." Paper presented at Evaluating Your Infectious Waste Disposal Contractors course, Department of Engineering Professional Development, University of Wisconsin–Madison (December 15, 1988).
13. U.S. Environmental Protection Agency. "Hospital Waste Combustion Study, Data Gathering Phase," final draft. Prepared by Radian Corp. (October 1987).
14. Brunner, C. R. "Incineration: Today's Hot Option for Waste Disposal." *Chem. Eng.* (October 12, 1987), pp. 96–106.
15. Tessitore, J. L. and F. L. Cross. "Incineration of Hospital Infectious Waste." *Poll. Eng.* (November 1988), pp. 83–88.
16. American Society of Mechanical Engineers. "Hazardous Waste Incineration A Resource Document" (New York: American Society of Mechanical Engineers, January, 1988).
17. Marks, C. H. "To Burn or Not to Burn: The Hospital's Modern-Day Dilemma." *Poll. Eng.* (November 1988), pp. 97–99.
18. Davis, J. and A. Fossa. "Hospital Waste Incineration: A New York State Perspective," in *Proceedings of the 1st National Symposium on Incineration of Infectious Waste*, Washington DC, May 5–6, 1988 (St. Louis, MO: Randolph, Breyer and Associates, Inc.)
19. U.S. Environmental Protection Agency, Midwest Research Institute. "Hospital Incinerator Operator Training Course," 3 vols. NTIS/PB89-189872 (March 1989).

CHAPTER 8

Other Treatment Technologies

In addition to steam sterilization and incineration—the principal techniques now used for treatment of infectious wastes—other technologies are also effective. Although none of these other techniques is used on a large scale in the United States at present, each one can be useful in special situations or for treating certain types of infectious waste.

Alternative technologies for treating infectious waste include the following:

- thermal inactivation/sterilization
- gas sterilization
- chemical disinfection
- irradiation
- discharge to the sanitary sewer system
- innovative technologies

Each of these alternatives is discussed briefly in this chapter in order to provide sufficient information so that you can evaluate the alternatives, understand the advantages and disadvantages of each, and decide on their feasibility for your particular needs and potential uses.

See Table 5.1 for a comparison of various infectious waste treatment technologies.

THERMAL INACTIVATION/STERILIZATION

Dry heat has long been used for the sterilization of medical supplies, equipment, and instruments.[1-3] More recently, this method of sterilization has been applied to the treatment of infectious wastes.

In the context of infectious waste treatment, thermal inactivation/sterilization refers to the treatment of waste with dry heat, i.e., without the addition of water, steam, or fire. For dry heat to sterilize waste, the waste must be maintained at a certain temperature so that the target microorganisms are exposed to that temperature for a minimum period of time.

Table 8.1 shows the range of times and temperatures required for dry heat sterilization under ideal conditions. These data are based on the conditions required to kill spores of *Bacillus subtilis* var. *niger* (*globigii*). Because ideal conditions are almost impossible to achieve when infectious waste is being treated, sterilizing cycles used in actual treatment operations are usually longer than those indicated in the table. A safety factor of two is generally used—that is, the waste is kept at treatment temperature for at least twice the length of time needed to kill the spores of the indicator microorganism.[1]

Table 8.1. Dry Heat Sterilization

Temperature		Spore Kill Time[a]
°C	°F	(hours)
121	250	6
140	285	3
150	300	2.5
160	320	2
170	340	1
180	356	0.5

Source: Draft Manual for Infectious Waste Management. U.S. EPA, SW-957, Washington, DC, September 1982 (Table 4-3, p. 4–25). Table was based on data from: E. Hanel, Jr., "Chemical Disinfection," in *Control of Biohazards in the Research Laboratory,* Course Manual, The Johns Hopkins University, School of Hygiene and Public Health, Baltimore, MD, 1981.

[a]In heat sterilization, exposure time for treatment is usually at least double the spore kill time.[1]

Dry heat is less efficient than moist heat as a sterilizing agent, and longer treatment times are necessary. (Compare Tables 6.1 and 8.1.) Because it is less efficient, this technique is also more expensive than steam sterilization, and cost can be an important factor when dry heat sterilization is being considered for the treatment of infectious waste.

Applicability

Despite its higher cost, dry heat treatment does have special applications. It is even the treatment of choice for certain wastes.

Wastewater

Certain types of infectious wastewater are routinely sterilized by heat before they are discharged into the sanitary sewer system or into a receiving stream such as a river. Examples of these wastes include some wastewaters from research and industrial laboratories and from pharmaceutical production. For these wastewaters, the treatment of choice is often thermal inactivation/sterilization.

Two different configurations of the treatment system are commonly used: a holding tank for batch treatment and a piping system for continuous treatment. Selection of the appropriate configuration depends on many factors that are specific to the particular use, such as the pattern of wastewater generation (batch or continuous), the quantities generated, and physical constraints.

When wastewater is treated in a tank, heat is applied by heat coils or an external steam jacket. Temperature of the contained wastewater must be monitored to be sure that the treatment temperature is reached and maintained for the necessary period of time. It is also important to mix the waste within the tank to keep the temperature uniform and to prevent temperature gradients and layering.

Alternatively, external steam jackets can be used to heat and thereby treat the wastewater as it flows through pipes between the point of generation (or a holding tank) and the point of discharge. This process is potentially much more efficient than treating wastewater in tanks because heat exchangers are usually part of the system. Heat exchangers serve two purposes: they cool the treated wastewater before it is discharged while using the recovered heat to heat the incoming wastewater. With this process, it is essential to control the flow of wastewater through the system at a rate that maintains the wastewater at treatment temperature for the necessary period of time. Such control requires monitoring the flow rate and the wastewater temperature as well as testing the effluent periodically for treatment effectiveness.

Sharps

Sharps can be sterilized by heating in a hot-air oven. Depending on the procedures that are used, this technique can also serve another purpose: it can destroy the syringes and render the sharps nonreusable. Sharps and syringes are placed together in the oven. The syringes melt at the high treatment temperature, and then, as the oven cools, the plastic solidifies around the sharps, forming a solid block. For the block to be handled safely, the sharps must be completely imbedded within the plastic.

Other Wastes

Other wastes, solid or liquid, can be treated effectively by dry heat sterilization in hot-air ovens. Such a procedure, however, is inefficient, time-consuming, and costly. Nevertheless, thermal inactivation/sterilization is an acceptable alternative technology for the treatment of infectious wastes.

Operation of a Hot-Air Sterilizer

Dry heat sterilization is usually done in a hot-air oven (for solid or liquid infectious wastes), in a tank (for bulk liquids), or in a piping system (for liquids). The typical operational cycle for a heat sterilizer consists of:

1. *warmup period*, when the temperature of the waste contained in the oven, tank, or piping is raised to sterilization temperature
2. *sterilization period*, when the temperature of the waste is maintained at the designated sterilization temperature for the time required for sterilization
3. *cooldown period*, when the temperature is lowered to allow handling, removal, and disposal of the treated waste

The length of time at treatment temperature is critical for achieving sterilization. You must determine this sterilization period when you are standardizing the operating conditions. The times needed for warmup and cooldown are then added to the sterilization period in order to establish the total length of the treatment cycle.

Temperature can be monitored using a temperature-measuring device such as a thermocouple that can be placed directly into the waste. In addition, a chart recorder is typically used to graph the temperature continually. Such a recorder is useful for quality control to check that the equipment functioned properly and to document the operational cycle.

Certain waste characteristics can impede thorough heating of the waste and thereby interfere with sterilization. Waste characteristics that are important in dry heat sterilization include mass or weight, heat capacity, heat conductivity, and barriers to heat convection.

Mass or Weight

Heat is absorbed by a waste in proportion to the mass or weight of the waste. The greater the mass/weight of the waste, the more energy is needed to heat that waste to treatment temperature. Very heavy wastes may require an extended warmup period.

Heat Capacity

The amount of heat required to raise the temperature of a material incrementally is called its heat capacity. Different materials have different heat capacities.

Therefore, types of waste may differ in the amount of heat needed to reach sterilization temperature, even when the weights of the loads are identical.

Heat Conductivity

Heat conduction, the movement of heat through an object, is a function of the heat conductivity of the material. The greater the heat conductivity of the waste, the more efficient the heating process is. Wastes that are poor heat conductors take longer to reach sterilization temperature, and the warmup periods must be longer. Poor heat conductivity is characteristic of wastes such as plastics, body tissues and animal carcasses, and animal bedding. The good heat conductivity of metal can be utilized to enhance heating and to reduce the warmup period through the use of metal pans and containers to hold the waste during thermal treatment.

Barriers to Heat Convection

Heat is also transferred to the waste by hot air convection. For efficient heating of waste during the warmup period, the hot air must be able to circulate freely throughout the chamber and, when possible, through the waste. This requires careful selection of waste containers (e.g., use of shallow metal pans and avoidance of plastic bags and tall narrow containers) as well as careful loading of the waste inside the sterilizer.

Quality Assurance

The biological indicator used for monitoring dry heat sterilization is the spore of *Bacillus subtilis* var. *niger* (*globigii*), the form of microorganism that is most difficult to kill using dry heat. For proper control and validity of testing, the biological indicators should be placed within the waste at different locations throughout the waste load.

The effectiveness of thermal treatment of liquid wastes can be monitored by using as target organisms the microorganisms that are present in the waste.

GAS STERILIZATION

Gas sterilization was also used originally to sterilize medical equipment and various medical and industrial products before use.[4-8] The practice of using gas for sterilization of supplies was extended more recently to its application for treatment of infectious wastes.

In the context of this chapter, gas sterilization is the treatment of infectious waste by exposing it to a sufficiently high concentration of a sterilizing gas under the required conditions for the designated treatment period. Ethylene oxide and formaldehyde are the sterilizing agents usually used in gas sterilization. Research is in progress on the use of hydrogen peroxide as the sterilant in gas sterilizers; if they become available, such units could also be used for treatment of waste in addition to their designated purpose of sterilizing medical supplies.

There is now substantial evidence that both ethylene oxide and formaldehyde are probable human carcinogens.[9-11] With both of these chemicals, there is also danger of exposure to gases that are slowly released from the waste after treatment has ended. The hazards that accompany the use of these chemicals impose strict constraints on gas sterilization operations and greatly limit the usefulness of this technology for treatment of infectious wastes.

The authors strongly discourage the use of ethylene oxide and formaldehyde for waste treatment when alternative treatment technologies are available. The use of ethylene oxide for waste treatment is not recommended,[12] and formaldehyde should be used only when it is the most appropriate treatment agent for the particular waste (and then used with great care).

If you are considering using gas sterilization for treating infectious waste, you must carefully evaluate the relative hazards of the waste and of the treatment operations. Do the benefits outweigh the hazards? If this technology is used, it is essential to minimize the potential for occupational exposure. This can be done by following proper operating procedures, using suitable personal protective equipment, and taking precautions to avoid exposure.

Because of the hazards (as well as the high costs) usually associated with gas sterilization, this technology is not, and should not be, used routinely or extensively for waste treatment. Nevertheless, there are special situations in which gas sterilization may be the preferred treatment method.

Ethylene Oxide

Ethylene oxide sterilizers are used in many hospitals to sterilize medical equipment and other items used in patient care that are spoiled or destroyed by heat or moisture and therefore cannot be sterilized by either steam or dry heat. The equipment is already installed on the hospital premises, and it would seem logical to use it for waste treatment also. However, although ethylene oxide may be essential for sterilizing some medical equipment and supplies, its use for treatment of *waste* is not recommended.[12] Besides, alternative treatment technologies are usually available.

Hazards

Extreme care is required if there is a possibility of worker exposure to ethylene oxide. Because of the toxicity and carcinogenicity of ethylene oxide,[9,10] the U.S. Department of Labor's Occupational Safety and Health Administration (or its state counterpart) requires that employees be protected from exposure to this chemical.[13] A gas detector must continually monitor the workplace atmosphere. Personal monitors are also sometimes required. The use of personal protective equipment such as respirators and gloves may be appropriate.

During treatment, ethylene oxide is absorbed by certain materials (such as rubber and plastic), and then the absorbed chemical is slowly released after treatment. To protect workers from exposure to the chemical due to this degassing process, the sterilization chamber is sometimes aerated to flush out released gas, or it may be vented for an extended period.

Release of ethylene oxide to the atmosphere is of increasing concern. Dilution and dispersal of the vented gas may be a means of disposal for some institutions. However, some states have adopted strict laws regulating release of toxic chemicals (especially carcinogens) to the atmosphere. Atmospheric release can be prevented by using gas scrubbers that capture and destroy the ethylene oxide.

Another problem with using ethylene oxide is that it is a very flammable and reactive gas. Gas cylinders should be stored and used away from heat and sources of ignition. The inventory of gas cylinders should be kept to a minimum. Obviously, the inventory would have to be larger if the gas were also used for waste treatment.

It is important to note that EPA lists ethylene oxide as an Extremely Hazardous Substance.[14] Because of this designation, institutions or facilities with more than 1000 lb of the chemical stored on site must notify the Local Emergency Planning Committee and the State Emergency Response Commission.

Applicability

Ethylene oxide sterilization of wastes is not recommended because the chemical is hazardous.[9,10] Waste treatment using ethylene oxide is also expensive, more so than other technologies. Use of this technology for infectious waste treatment should be limited to special situations such as the temporary unavailability of other treatment equipment and the need to sterilize infectious waste that otherwise would be accumulating untreated.

Formaldehyde

Formaldehyde has long been used as the gaseous agent for sterilizing spaces and surfaces such as rooms, buildings, and ventilation systems.[7] In the context of infectious waste treatment, formaldehyde sterilization is important because it can be used in special situations for which no other treatment technique is equally effective or suitable.

Gas sterilization using formaldehyde is usually the method of choice for treating waste items that must be sterilized in situ. The hazards present in the waste often warrant sterilizing it in place, where the waste is generated, before it is moved. For example, one application is formaldehyde sterilization of high-efficiency particulate air (HEPA) filters in biological safety cabinets.

Factors affecting the success of formaldehyde sterilization include concentration of gas, relative humidity, temperature, and contact time. The area that is being treated must be sealed off to minimize the risk of occupational exposure.

Formaldehyde is hazardous, and there is substantial evidence that it is a human carcinogen.[9,11] There are OSHA regulations to protect employees from occupational exposure to formaldehyde.[15] Only persons who are properly trained in the use of formaldehyde as a gaseous sterilant should perform formaldehyde sterilization procedures. Additional precautions to avoid exposure must be taken after the treatment cycle is ended because formaldehyde frequently leaves a residue and degassing continues for some time.

EPA lists formaldehyde as an Extremely Hazardous Substance.[14] Therefore, institutions or facilities with more than 500 lb of the chemical stored onsite must notify the Local Emergency Planning Committee and the State Emergency Response Commission.

CHEMICAL DISINFECTION

Chemical disinfection is used routinely in medical care[16] for cleaning certain instruments and supplies,[17] for surgical scrubs,[18] and for general cleaning of floors, walls, and furniture. As applied to infectious waste treatment, chemical disinfection is treatment of the waste by addition of chemicals that kill or inactivate the infectious agents. This type of treatment usually results in disinfection rather than sterilization.

Chemical disinfection is most suitable for use in treatment of liquid wastes. Chemical disinfection can also be used to treat solid infectious waste that is shredded before or during treatment in, for example, a hammermill. With intact solid infectious wastes, chemical treatment provides only surface disinfection.

Operation Considerations

In order to treat infectious wastes effectively by chemical disinfection, it is essential:

- to use the proper disinfectant
- to add a sufficient quantity of the chemical
- to allow sufficient contact time
- to control other conditions as necessary

Type of Disinfectant

Various disinfectants are effective in killing or inactivating microorganisms. Some disinfectants are effective only against specific types of microorganisms, but few disinfectants can be used for all types of infectious agents. (See Table 8.2.) Therefore, in order to select a disinfectant, it is essential to know which

Table 8.2.   Selected Chemical Disinfectants[a]

	Chlorine Compounds	Iodophor	Alcohols[b]	Formaldehyde	Glutaraldehyde
Inactivates					
Vegetative bacteria	yes	yes	yes	yes	yes
Lipoviruses	yes	yes	yes	yes	yes
Nonlipid viruses	yes	yes	[c]	yes	yes
Bacterial spores	yes	yes	no	yes	yes
Treatment requirements					
Use dilution	500 ppm[d]	25–1600 ppm[d]	70–85%	0.2–8.0%	2%
Contact time, min.					
Lipovirus	10	10	10	10	10
Broad spectrum	30	30	not effective	30	30
Important characteristics					
Effective shelf life >1 week[e]	no	yes	yes	yes	yes
Corrosive	yes	yes	no	no	no
Flammable	no	no	yes	no	no
Explosion potential	none	none	none	none	none
Inactivated by organic matter	yes	yes	no	no	no
Skin irritant	yes	yes	no	yes	yes
Eye irritant	yes	yes	yes	yes	yes
Respiratory irritant	yes	no	no	no	no
Toxic[f]	yes	yes	yes	yes	yes
Applicability					
Waste liquids	yes	no	no	no	no
Equipment surface decontamination	yes	yes	yes	yes	yes

[a]Adapted from Table 4-6 in *Draft Manual for Infectious Waste Management.* U.S. EPA, SW-957, Washington, DC, September 1982, pp. 4-35 to 4-36. (Table was adapted from *Laboratory Safety Monograph. A Supplement to the NIH Guidelines for Recombinant DNA Research.* National Institutes of Health, Office of Research Safety, National Cancer Institute, and the Special Committee of Safety and Health Experts, Bethesda, MD, January 1979, pp. 104–105.)
[b]Ethyl and isopropyl alcohols.
[c]Results are variable, depending on the virus.
[d]Concentration of available halogen.
[e]When protected from light and air.
[f]For exposure by skin or mouth or both. Refer to manufacturer's literature or Merck Index.

microorganism is the target of the treatment procedure, to understand the biology of the target microorganism(s), and to characterize the nature of the waste.

Disinfectants include chlorine compounds,[19] alcohols,[20] phenolic compounds,[21] iodine compounds,[22] quaternary ammonium compounds,[23] formaldehyde,[7] and glutaraldehyde.[24] Most of these are usually used for disinfection of contaminated surfaces and medical equipment, while formaldehyde is commonly used to preserve and disinfect pathological specimens. Chlorine compounds (primarily sodium hypochlorite) constitute the only type of disinfectant that is used routinely to disinfect infectious wastes.

Quantity of Disinfectant

The quantity and concentration of the disinfectant are important factors for ensuring the effectiveness of chemical treatment. Sufficient disinfectant must be added to react with all the infectious agents present in the waste.

The quantity of disinfectant that is needed is a function of the chemical used, its concentration, and the degree of contamination of the waste. Another important factor is the amount of proteinaceous material present in the waste, because such material binds the disinfectant and prevents it from reacting with the infectious agents. Therefore, disinfectant must always be added in excess so that there is enough of the chemical available to ensure treatment effectiveness.

Contact Time

As with every other type of treatment, the duration of treatment—that is, the contact time—is an important factor for achieving effective chemical treatment. In order for disinfection to occur, there must be enough chemical available to react with the infectious agents that are present in the waste, and sufficient time must be allowed for this reaction to take place.

Other Factors

Various other factors are relevant to chemical treatment of infectious wastes. These factors include pH, temperature, and mixing.

The pH of the reaction mixture might enhance or impede the disinfection process. For rapid and thorough disinfection, one must use the appropriate pH—that is, the pH at which the kill rate is highest and fastest for the particular disinfectant acting on the target microorganism(s).

Like pH, temperature can also affect the success of chemical treatment. Changes in temperature might increase or decrease the effectiveness and rapidity of treatment.

Mixing requirements are primarily a function of the type and the quantity of waste. In small quantities of aqueous solutions, the disinfectant can diffuse rapidly throughout the waste. Mixing is required for large quantities of liquid waste in order to ensure sufficient and effective contact between the chemical disinfectant and the infectious agents. Shredded solid wastes must also be mixed after addition of disinfectant because diffusion through such material is slow, and treatment would be inadequate without mixing unless contact time were greatly prolonged.

Applicability

The principal use of chemical disinfection is in treating liquid infectious wastes. Effective chemical treatment of solid wastes is much more difficult to achieve.

Liquid Infectious Wastes

Liquid infectious wastes such as discarded culture media are usually well characterized in terms of chemical constituents, pH, and microbial load. Composition of these wastes is often uniform through successive batches, and it is relatively easy to determine required treatment conditions and to validate treatment effectiveness. Even when the wastes are not uniform in composition, one can easily test for treatment effectiveness by assaying the treated waste for the target microorganism(s).

Solid Infectious Wastes

With the hammermill/chemical disinfection technology, the solid infectious waste is shredded and ground as part of the treatment process. Wet grinding is usually used, with the disinfectant solution (rather than water) wetting the waste. Some treatment systems are designed to treat sharps, pathological wastes, and laboratory wastes. Other systems are capable of handling general infectious wastes contained in plastic bags, bottles, and boxes.[25] At the end of the process train, the slurry resulting from treatment is drained—the solids are collected for landfilling while the liquid is discharged to the sewer system.

Several occupational safety and health considerations are relevant to this technology. It is essential that the shredding and grinding be conducted under negative pressure to prevent release of potentially infectious aerosols and worker exposure. Personal protective equipment should be used to protect against exposure to the waste and to the disinfectant, and hearing protection is needed for employees working near noisy equipment.

Regulatory Considerations

Wastewater from chemical treatment that is discharged into the sanitary sewer system must be in compliance with federal, state, and local regulations. These regulations impose certain limitations on the chemical constituents and other parameters of wastewater. Potential problems are the presence of the chemical disinfectant and the increased organic content of the wastewater. In addition, when solids are ground up as part of the treatment process, the high concentration of suspended solids in the wastewater could be a problem.

Quality Assurance

A testing program is needed when chemical treatment is first used to determine the treatment conditions necessary to ensure effective treatment of the waste. Load standardization can be based on a single worst-case load or on a number of clearly defined standard loads. See Chapter 5 for a detailed discussion of these two options and the establishment of treatment conditions and operating protocols.

With liquid wastes, the best method of testing treatment effectiveness is to assay the treated waste for the known target microorganism(s). Testing is much more complicated when solid infectious wastes are treated by grinding and chemical disinfection. Research now in progress may provide testing protocols that can be used for testing treatment effectiveness and for quality assurance.

IRRADIATION

Radiation is used for sterilizing certain products and supplies.[26-29] In this country, radiation is seldom used to sterilize infectious wastes. Application of this technology to treatment of infectious waste is limited because of high costs, the need for extensive protective equipment, the requirement for highly trained operating personnel, and problems with disposal of the radioactive source.

Ultraviolet light cannot penetrate material to any depth; therefore, its use is limited to the sterilization of surfaces. One good use for ultraviolet light in waste treatment is the sterilization of sheets of paper. This is a very specialized application that is certainly not efficient in routine treatment of infectious waste.

Gamma rays from the radioisotope cobalt-60 do penetrate to a greater depth. Gamma ray irradiation is therefore a technology that is applicable to infectious waste sterilization. However, it is now rarely used for waste treatment. It is impossible to predict if or when this technology might come into general use.

DISCHARGE TO THE SANITARY SEWER SYSTEM

Some experts advocate the discharge of untreated liquid infectious waste to the sanitary sewer system, maintaining that the wastewater treatment system provides the best technology for treating biological wastes. This may be true, but there is an important caveat: the waste must receive secondary treatment at the wastewater treatment plant. Unfortunately, not all sewer systems provide secondary treatment.

Another problem is that some systems (including those of some major metropolitan areas) are combined sanitary/storm sewers that receive both wastewater and the runoff from storms. In the event of major storms, the capacity of the wastewater treatment plant is frequently insufficient to handle the total inflow, and the plant must then be bypassed, with untreated sewage being discharged directly into the receiving stream.

Liquid infectious waste can be disposed of into the sanitary sewer system for treatment at the wastewater treatment plant if the system does provide secondary treatment and if it is not a combined sanitary/storm sewer system. Untreated infectious wastes should not be placed into a combined sanitary/storm sewer system.

When this technique is used, precautions must be taken to prevent worker exposure. These include use of procedures and equipment that minimize aerosol generation,* use of personal protective equipment such as face shields (or masks and goggles), gowns, and gloves, restriction of disposal to designated sinks and drains, and disinfection of the sinks and drains before any maintenance work is done.

Infectious waste disposed of to the sanitary sewer system must meet state and local regulations on wastewater quality. These regulatory requirements usually impose limits on a variety of constituents and parameters including chemicals, pH, organic material (biochemical oxygen demand or BOD), and total suspended solids.

INNOVATIVE TECHNOLOGIES

In the interests of promoting technological development, we must keep an open mind regarding innovative technologies for treatment of infectious wastes. New techniques or equipment may some day provide treatment alternatives. Although these new techniques and technologies may not fit precisely into one of the treatment categories that are discussed in this chapter, this must not make them automatically unacceptable.

In evaluating treatment alternatives, old as well as new, we must take the pragmatic (rather than the dogmatic) approach. The primary criterion must be: DOES IT WORK? If the answer is "yes," then that treatment technique is acceptable, once operating, QA/QC, and safety procedures have been developed and validated.

---

*Unpublished data indicate that microbial contamination is present when there is visible splash and splatter. Splash and splatter can be reduced by minimizing the distance between the pour spout and the water level (it should be less than 10–12 inches) and by using a container with a good pour spout.[30]

REFERENCES

1. Perkins, J. J. "Dry Heat Sterilization," in *Principles and Methods of Sterilization in Health Sciences*, 2nd ed. (Springfield, IL: Charles C. Thomas, 1969), pp. 286–311.
2. Pflug, I. J. "Heat Sterilization," in *Industrial Sterilization*, G. B. Phillips and W. S. Miller, Eds. (Durham, NC: Duke University Press, B-D Technology Series, 1973), pp. 239–282.
3. Joslyn, L. J. "Sterilization by Heat," in *Disinfection, Sterilization, and Preservation*, 3rd ed., S. S. Block, Ed. (Philadelphia: Lea & Febiger, 1983), pp. 3–46.
4. Perkins, J. J. "Sterilization of Medical and Surgical Supplies with Ethylene Oxide," in *Principles and Methods of Sterilization in Health Sciences*, 2nd ed. (Springfield, IL: Charles C. Thomas, 1969), pp. 501–530.
5. Ernst, R. R. "Ethylene Oxide Gaseous Sterilization for Industrial Applications," in *Industrial Sterilization*, G. B. Phillips and W. S. Miller, Eds. (Durham, NC: Duke University Press, B-D Technology Series, 1973), pp. 181–208.
6. Hess, H., L. Geller, and X. Buehlmann. "Ethylene Oxide Treatment of Naturally Contaminated Materials," in *Industrial Sterilization*, G. B. Phillips and W. S. Miller, Eds. (Durham, NC: Duke University Press, B-D Technology Series, 1973), pp. 283–296.
7. Tulis, J. J. "Formaldehyde Gas as a Sterilant," in *Industrial Sterilization*, G. B. Phillips and W. S. Miller, Eds. (Durham, NC: Duke University Press, B-D Technology Series, 1973), pp. 209–238.
8. Caputo, R. A., and T. E. Odlaug. "Sterilization with Ethylene Oxide and Other Gases," in *Disinfection, Sterilization, and Preservation*, 3rd ed., S. S. Block, Ed. (Philadelphia: Lea & Febiger, 1983), pp. 47–64.
9. U.S. Environmental Protection Agency, Carcinogen Assessment Group (CAG). "Relative Carcinogenic Potencies Among 55 Chemicals Evaluated by CAG as Suspect Carcinogens. Mutagenicity and Carcinogenicity Assessment of 1,3-Butadiene, Final Report." EPA/600/8-85/004F. Washington, DC (September 1985).
10. U.S. Department of Labor, Occupational Safety and Health Administration. "Occupational Exposure to Ethylene Oxide; Final Standard." *Federal Register* 49(122):25734–25809 (June 22, 1984).
11. U.S. Department of Labor, Occupational Safety and Health Administration. "Occupational Exposure to Formaldehyde; Final Rule." *Federal Register* 52(233): 46168–46312 (December 4, 1987).
12. U.S. Environmental Protection Agency. *EPA Guide for Infectious Waste Management*. EPA/530-SW-86-014. Washington, DC (May 1986).
13. *Code of Federal Regulations*, Title 29, Part 1910.1047, "Ethylene Oxide."
14. *Code of Federal Regulations*, Title 40, Part 355, Appendix A. "The List of Extremely Hazardous Substances and Their Threshold Planning Quantities."
15. *Code of Federal Regulations*, Title 29, Part 1910.1048, "Formaldehyde."
16. Perkins, J. J. "Chemical Disinfection," in *Principles and Methods of Sterilization in Health Sciences*, 2nd ed. (Springfield, IL: Charles C. Thomas, 1969), pp. 327–344.
17. Favero, M. S. "Chemical Disinfection of Medical and Surgical Materials," in *Disinfection, Sterilization, and Preservation*, 3rd ed., S. S. Block, Ed. (Philadelphia: Lea & Febiger, 1983), pp. 469–492.

18. Altemeier, W. A. "Surgical Antiseptics," in *Disinfection, Sterilization, and Preservation*, 3rd ed., S. S. Block, Ed. (Philadelphia: Lea & Febiger, 1983), pp. 493–504

19. Dychdala, G. R. "Chlorine and Chlorine Compounds," in *Disinfection, Sterilization, and Preservation*, 3rd ed., S. S. Block, Ed. (Philadelphia: Lea & Febiger, 1983), pp. 157–182.

20. Morton, H. E. "Alcohols" in *Disinfection, Sterilization, and Preservation*, 3rd ed., S. S. Block, Ed. (Philadelphia: Lea & Febiger, 1983), pp. 225–239.

21. Prindle, R. F. "Phenolic Compounds," in *Disinfection, Sterilization, and Preservation*, 3rd ed., S. S. Block, Ed. (Philadelphia: Lea & Febiger, 1983), pp. 197–224.

22. Gottardi, W. "Iodine and Iodine Compounds," in *Disinfection, Sterilization, and Preservation*, 3rd ed., S. S. Block, Ed. (Philadelphia: Lea & Febiger, 1983), pp. 183–196.

23. Petrocci, A. N. "Surface-Active Agents: Quaternary Ammonium Compounds," in *Disinfection, Sterilization, and Preservation*, 3rd ed., S. S. Block, Ed. (Philadelphia: Lea & Febiger, 1983), pp. 309–329.

24. Scott, E. M., and S. P. Gorman. "Sterilization with Glutaraldehyde," in *Disinfection, Sterilization, and Preservation*, 3rd ed., S. S. Block, Ed. (Philadelphia: Lea & Febiger, 1983), pp. 65–88.

25. Denys, G. A. "Microbiological Evaluation of the Medical SafeTEC Mechanical/Chemical Infectious Waste Disposal System." Paper presented at the 89th Annual Meeting of the American Society for Microbiology, New Orleans (May 14–18, 1989). *Program of the 89th Annual Meeting of ASM*, p. 101, Abstract #Q 57.

26. Plester, D. W. "The Effects of Radiation Sterilization on Plastics," in *Industrial Sterilization*, G. B. Phillips and W. S. Miller, Eds. (Durham, NC: Duke University Press, B-D Technical Series, 1973), pp. 141–152.

27. Eymery, R. "Design of Radiation Sterilization Facilities," in *Industrial Sterilization*, G. B. Phillips and W. S. Miller, Eds. (Durham, NC: Duke University Press, B-D Technical Series, 1973), pp. 153–179.

28. Silverman, G. J. "Sterilization by Ionizing Irradiation," in *Disinfection, Sterilization, and Preservation*, 3rd ed., S. S. Block, Ed. (Philadelphia: Lea & Febiger, 1983), pp. 89–105.

29. Shechmeister, I. L. "Sterilization by Ultraviolet Irradiation," in *Disinfection, Sterilization, and Preservation*, 3rd ed., S. S. Block, Ed. (Philadelphia: Lea & Febiger, 1983), pp. 106–124.

30. Rutala, W. Director of Statewide Infection Control Program and Associate Research Professor, Department of Medicine, Division of Infectious Diseases, University of North Carolina, Chapel Hill, NC. Personal communication to Judith G. Gordon (September 6, 1989).

CHAPTER 9

Disposal of Treated Waste

When infectious waste has been properly and effectively treated, the waste is no longer infectious. The treated waste then poses essentially no risk, and it can be handled in the same way as ordinary solid waste.

There are, however, two exceptions to this generalization.

1. For certain types of waste, such as sharps and pathological wastes, additional processing before disposal may be warranted or even necessary, depending on the type of treatment used. (See sections below on these two types of waste.)
2. If other hazards in addition to infectiousness were present in the waste (e.g., chemical toxicity or radioactivity), there must be additional treatment before disposal and/or special handling and disposal. (Management of wastes with multiple hazards is discussed in Chapter 13.)

There are two disposal options for treated infectious waste: the sanitary landfill and the sanitary sewer. Regardless of the disposal method used, a quality assurance/quality control program is necessary to ensure that only treated waste is designated for disposal.

LANDFILL DISPOSAL

Burial in a sanitary landfill is often considered the best disposal option for general solid waste (that is, for noninfectious waste). Liquid wastes should not be placed into landfills; these wastes are best disposed of by discharge to the sanitary sewer system.

Landfills are usually used for the disposal of treated infectious waste and the residue from treatment. Treated wastes are those that were treated by, for example, steam sterilization. Treatment residues include incinerator ash and ground-up wastes (that is, wastes that were ground during or after treatment).

Landfill disposal is the preferred option for most treated waste and treatment residues. There may be problems, however, with the landfilling of certain types of waste that were treated by steam sterilization. These particular wastes are sharps, pathological wastes, and red bags. (See discussion below.)

Incinerator Ash

Incinerator ash includes both bottom ash and flyash. Bottom ash is the non-combustible component of the waste that remains in the incinerator chamber after burndown. Flyash consists of the lighter particulates that are entrained in the flow of combustion gases. Some air pollution control devices are effective in collecting the flyash.

Incinerator ash is generally disposed of by burial in landfills. This is usually acceptable, providing that the ash is neither toxic nor radioactive. Flyash from medical waste incineration is sometimes hazardous by the EP toxicity test; in this case, it must be disposed of in a hazardous waste landfill rather than a sanitary landfill.

Sharps

In addition to potential infectiousness, sharps have an inherent physical hazard—the potential for causing cuts, scratches, and puncture wounds. This physical hazard can persist even when sharps are sterilized. (See below.) When intact sharps are sent to a landfill for disposal, workers at the landfill are at risk for injury by the sharps. Workers at risk include:

- waste handlers, who can be injured by sharps that are not properly contained
- equipment operators, who can be injured by sharps that are released from containment during burial and compaction operations at the landfill
- equipment maintenance workers, who can be injured by sharps caught in the treads and wheels of compaction equipment and other landfill vehicles

Furthermore, regulations in some states require that all needles and syringes be rendered ''nonreusable'' before disposal. These regulations were issued in an attempt to control drug abuse by preventing unauthorized access to usable needles and syringes.

It is important to consider these factors when you evaluate your sharps management system, select a new system, or change procedures. Your system should address all aspects of sharps management: their potential infectiousness, their physical hazard, the possibility of reuse, and regulatory requirements.

With proper incineration, the metal in disposable needles becomes friable and flaky. The needles are thereby effectively destroyed, and there is little risk of physical injury or possibility of reuse. Therefore, there is essentially no problem associated with the disposal of incinerated needles in landfills.

When needles are treated by steam sterilization, however, they remain intact and reusable; the physical hazard persists because steam does not affect the physical characteristics of these wastes. These needles are a hazard and a cause for concern. Furthermore, in states with regulations requiring that needles and syringes be rendered nonreusable before disposal, steam sterilization alone is not sufficient. An additional processing step is usually necessary to render the needles and syringes nonreusable. Acceptable alternatives include imbedding in plastic, grinding, and compaction.* The imbedded, ground, or compacted sharps can then be safely disposed of in a landfill—they cannot cause injury or be reused.

In consideration of all these factors, it is best to manage sharps so that they are not intact when they leave the facility. All needles should be destroyed or similarly processed before they are sent to the landfill. The best way to eliminate the physical hazard of needles, to meet regulatory requirements, and to avoid liability is to establish a sharps management program that ensures that no intact needles are sent to the landfill for disposal.

Pathological Wastes

Because of aesthetic considerations, body parts that are recognizable should not be placed into landfills. When pathological wastes are incinerated and there is complete burndown of the waste, the ash can be disposed of in a landfill.

Sometimes pathological wastes are steam sterilized because of their potential infectiousness. Pathological wastes treated in this way require additional processing because of the above-mentioned aesthetic considerations. Several options are available. These wastes can be transferred to a mortician for burial or incineration in a crematorium, incinerated in a pathological incinerator, or ground up and flushed to the sanitary sewer system (if this meets the requirements of the local sewer authority).

Red Bags

When red plastic bags full of treated waste arrive at a landfill for disposal, there is an immediate problem that is primarily one of perception. The waste may have been properly treated by steam sterilization, but the landfill worker

---

*Some types of equipment provide steam sterilization followed by additional processing as an integral part of the treatment system.

sees the red bag as a warning that the contents are dangerous. Solutions to this dilemma are not simple.

At present, practically all red bags retain their red color after steam sterilization. Many brands do provide some color change brought about by exposure to steam, but this usually involves only the appearance of some black or brown words or markings on the red bag. These color changes are not readily discernible to haulers and landfill workers, who tend to see only the *red bag*.

Because of this problem, a better container for infectious waste that is to be steam sterilized is the red plastic bag that crumples on exposure to steam. When this type of plastic bag is used, the bag practically disintegrates and it is readily apparent that the waste has been treated. Of course, use of a bag that crumples creates other problems—a secondary container such as a paper bag or a pan is then essential to contain the waste within the sterilizer. Nevertheless, use of such a bag eliminates the situation in which the hauler or landfill worker sees an intact red bag and automatically assumes that it contains infectious wastes.

Alternatively, the waste generator may be able to convince the hauler and the landfill operator that a strong quality assurance/quality control program is in place and functioning. This program ensures that all red bags are steam sterilized, that the treatment is proper and sufficient, and that only treated waste leaves the premises for disposal at the landfill. Under these circumstances, the waste hauler may be reassured and may be willing to pick up treated red bags for disposal at a landfill.

DISCHARGE TO THE SANITARY SEWER SYSTEM

Liquid wastes are unsuitable for disposal in a landfill, and incineration of aqueous wastes is fuel-intensive and inefficient. Therefore, treated waste that is liquid or semiliquid is best disposed of by discharge to the sanitary sewer system, provided that such discharge meets the requirements of the local sewer authorities for wastewater constituents and characteristics. These requirements may include limits on various parameters including chemicals, organic material (BOD), pH, total suspended solids, and temperature.

Liquid wastes suitable for sewer disposal are infectious wastes that were treated prior to their discharge into the sewer system. (See Chapter 8 for a discussion of the discharge of untreated infectious waste to the sanitary sewer system.) This includes wastes that were steam sterilized, those that were treated chemically, and those that received heat treatment.

Semiliquids are usually formed when wastes are ground up during or after treatment, with the semiliquid product being discharged directly to the sewer system. Sharps and pathological wastes are sometimes treated this way. In addition, certain infectious waste treatment systems pass the waste through a hammermill or grinder while it is being treated chemically; although the resultant slurry is drained on a conveyor belt, some semiliquid waste enters the sewer system.

CHAPTER 10

Minimizing Infectious Waste

Many terms are used to describe waste minimization efforts: reduction, abatement, prevention, avoidance, elimination, and source reduction. As the Office of Technology Assessment notes, no standard definition exists for any of these terms. Each is defined differently by each user.[1] This chapter uses the term "waste minimization" because it is the term that the EPA uses most often in reference to preventing the generation of waste by any of the methods described in Table 10.1.

Many definitions of waste minimization also include treatment methods that reduce or eliminate the hazard of a waste before its disposal. Steam sterilization, incineration, and other treatment methods for infectious waste are discussed in detail in Chapters 6, 7, and 8, respectively. Interestingly, steam sterilization can also be used to reduce waste volume by facilitating the reuse of supplies, and incineration has the advantage of being able to recover energy from infectious and medical waste: both treatment methods minimize waste in more than one way.

WHY WASTE MINIMIZATION?

The impetus for waste minimization usually stems from the need to control or reduce the cost of waste management, the need to preserve total landfill capacity, or a desire by regulators to reduce environmental impacts. When the high cost of infectious waste management came to the attention of administrators at the University of North Carolina, they reviewed infectious waste management practices for ways to reduce costs.[2] Increasing costs are not uncommon for infectious and medical waste. Increases may be due to new or more stringent regulations or simply higher fees charged by a hauler or landfill operator. As mentioned

Table 10.1.   Waste Minimization Methods[a]

Reducing the amount of materials used
Purchasing constraints
Substitution to reusable supplies
Substitution to less wasteful or less hazardous supplies
Other procedural changes

Reducing the amount of waste generated
Source separation
Waste segregation
Other procedural changes

Recycling and reuse
Steam sterilization
Gas sterilization
Other decontamination methods (chemical or radiation)

Volume reduction techniques
Incineration (also reduces weight)
Steam sterilization (minimal)
Compaction
Shredding, milling, crushing

Energy recovery techniques
Incineration with boiler

[a]See text for most appropriate methods for infectious waste.

in Chapter 1, these trends and rising disposal costs are likely to continue. In response, facility managers naturally ask if the volume of infectious and medical waste can be reduced. It can.

Interest in waste minimization is also frequently the result of restricted access to disposal routes. In 1978, for example, the University of Minnesota ran into problems with its hospital waste incinerator that spurred them to consider waste minimization options.[3] Today, new regulations are requiring medical facilities to abandon their previous methods of infectious waste disposal. Incinerators that cannot meet the new standards are closing. Landfills that previously accepted untreated infectious waste (often unknowingly) are on the lookout for red bags and refusing them. In many states, the definition of infectious waste is broadening—and waste volumes are ballooning with the stroke of a pen. In many rural areas, commercial disposal firms are unavailable or transport costs are excessively high.

An often-overlooked benefit of waste minimization is its reduction in occupational and environmental risk. Less waste results in less handling, smaller chance of exposure, lower incinerator emissions, and reduced possibility of a release. Also, when source separation is ignored and normal trash is haphazardly discarded with infectious waste, trash handlers and other staff start handling red bags as if they were normal trash; employees become complacent about the occupational hazards of infectious waste. Waste minimization lowers political risks too; at a minimum, such efforts demonstrate to the community that the institution is serious about reducing its environmental impact.

For a medical facility looking at waste management strategies for a solution to these problems, waste minimization is as much a part of waste management as commercial disposal. Chapter 1 explains that the most successful waste management plans incorporate a diversity of methods. Indeed, waste minimization is an essential component of all institutional waste management plans as a commitment to managing institutional risks. This chapter discusses several waste minimization options for infectious and medical waste. With the exception of source separation, which should be practiced by all generators, not all options are appropriate for all institutions.

DISPOSABLES

Ours is a throw-away society, and medical institutions are no exception. Many hospital engineers cite their experience that over the years, the per-bed total waste generation in hospitals has grown ever larger. People prefer the convenience of disposable items because of their newness, ease of disposal, and ready availability.

Disposables vs Reusables

Historically, hospitals have used many types of reusable supplies that were cleaned and sterilized between use. Autoclaves and gas sterilizers are still used by medical facilities for such purposes, but to a much lesser extent. For more than a decade, multiple-use (reusable) products have been increasingly replaced by single-use, disposable supplies.[4] Examples include the increased use of disposable needles and syringes and of disposable paper products in place of launderable cloth items. Manufacturers and institutions have cited that switching to single-use items represents a savings in reprocessing costs. Infection control officers have suggested that the use of fewer multiple-use items may reduce the risk of nosocomial disease. However, few decisions to replace reusable supplies have considered the added waste disposal costs of the single-use items, or the risks of waste management.

Lessons Learned at the University of Minnesota

A study at the University of Minnesota[3] carefully explored the possibilities of reversing the trend toward single-use disposable items. Lists of supplies were reviewed, policies defining infectious waste were examined, and external and internal disposal costs were calculated. As a result, several types of disposable single-use supplies were identified as candidates for replacement by reusable supplies. The potential benefit of these changes was then examined. The researchers found that the system for reusable supplies, including quality control of cleaning and sterilization, administration of central supply, storage space, and labor, is expensive, but concluded that ''the replacement of selected single-use disposable products with reusable products will not significantly reduce the total weight and disposal costs of hospital solid waste.'' Beneficial results of the study included a revised purchasing procedure that added disposal costs to the purchase price of single-use items, which led to a more accurate comparison when purchasing decisions came down to disposable or reusable supplies.

The study may have had one flaw. It noted that for many procedures, the possibility of contaminating an item with an infectious agent is remote; for certain items, some may become infectious waste, but most do not. However, the study failed to identify those items that, by their nature of use, nearly always are discarded, thereby becoming infectious waste. It is the replacement of these items with reusable supplies that is likely to result in significant waste minimization.

Changing Practices to Reduce Disposables

Although central purchasing controls to achieve waste minimization may be appropriate in some cases, the authors are comfortable with allowing some flexibility in purchasing practices rather than establishing rigid acquisition controls. A close examination of practices that generate infectious waste may disclose procedures in which fewer disposable items can be used or single-use supplies can be eliminated. Once the problems and costs of waste disposal, along with possible solutions, have been explained to them, employees tend not only to take up the crusade but often to discover other creative solutions.

SOURCE SEPARATION

Source separation is the most effective way to reduce the amount of infectious waste generated in a medical facility. Many locations where infectious waste is generated have separate collection containers for normal trash and infectious waste. Source separation is simply the practice of keeping normal trash out of the infectious and medical waste stream and vice versa. It sounds easy, but it isn't.

Source separation can never be absolutely efficient. A tour of waste collection bins or a visit to the infectious waste incinerator will likely reveal many instances of improper materials being disposed of with infectious waste. The goal of source separation is to keep normal trash and infectious waste separate to the maximum extent practicable. Toward this end, it makes sense to equip office areas with only normal trash bins and isolation rooms with only infectious waste collection containers. In most cases, however, two waste collection bins must be placed in areas that generate infectious waste. One configuration that works well is to put the infectious waste container side by side with the normal trash container.

To avoid confusion, the bins used for each type of waste should be clearly identified and distinct from one another. Color coding of the containers or using differently shaped containers are excellent ways to ingrain the different disposal routes. The biohazard symbol should also be prominently displayed on the infectious waste bin; its distinctiveness helps source separation become a subconscious habit.

In some cases, infectious waste is defined by the environment in which the material was used. For example, some hospitals consider all wastes coming from operating rooms, isolation patients, and dialysis centers to be infectious. If indeed this policy is appropriate, normal trash bins can be placed just outside these areas to encourage staff to unwrap supplies before entering these areas and dispose of packing materials as normal trash.

Inattention to source separation mistakes gives a message to staff that the institution doesn't really care about infectious waste management—the opposite employee attitude is needed for successful waste minimization. Institutional goals must be clear and consistently evidenced in daily practices.

Costs

Cost control is a powerful argument for source separation. Infectious waste is significantly more expensive to dispose of than normal trash.[3,4] In addition, the amount of regulated waste (i.e., infectious or medical waste as defined by federal or state regulations) is minor compared to the amount of normal trash.[3,5] Under most laws, the regulation of a waste is based on the waste component having the greatest risk. This means that a small amount of infectious waste placed in a load of normal trash can necessitate special disposal procedures for the entire load.

As explained in Chapter 17, the cost of waste management is difficult to determine. Offsite disposal costs are typically the focus of such analysis, but onsite handling and storage are usually more important expenses. Few analyses of disposal costs include administrative and operational overhead costs. The University of Minnesota study[3] found that internal (onsite) costs accounted for 63% of the total waste management cost. Since onsite costs for normal trash and infectious waste can be similar (both require handling and storage; labor is the most significant expense), differences in offsite disposal fees may have a minor effect on the total waste disposal bill. Source separation efficiency can vary so widely between institutions and departments, however, that better separation can result in dramatic cost savings.

Training Costs of Source Separation

Source separation requires effective and continuing employee education. The University of Minnesota found that ''an in-hospital education program [was an] absolute necessity in order for procedural changes to be accepted and followed by hospital personnel.''[3] Employee training should:

- Define infectious waste and outline the importance of regulations, policies, and methods and environments in which supplies are used. This is confusing to many employees because infectiousness isn't obvious. Also, how a product is used usually determines if it becomes infectious waste; except for sharps, the nature of a particular product doesn't determine its disposal route.
- Explain institution policies and procedures for waste generation and source separation. When discussing these with employees, solicit their views on the practicality of the written procedures. Also, identify those activities for which procedures are still unclear, incomplete, or nonexistent.
- Heighten employee awareness of the problems of infectious and medical waste management. Begin by assessing the staff's attitudes. Explain the costs and risks of waste management. Make sure employees understand the rationale behind waste management procedures. Many people today have a strong safety and environmental ethic; identifying those values with institutional goals will result in greater cooperation. In many cases, a voicing of employee ideas is the greatest benefit of these sessions.

The Importance of Defining Infectious Waste

Federal, state, and local laws specify wastes that must be considered infectious. Beyond these minimum standards, an institution may adopt guidelines and policies that broaden its definition of infectious waste. Such discretionary wastes can be a significant portion of the institution's infectious waste stream. Thus, the amount of infectious waste generated by an institution is largely determined by whether ''infectious waste'' is defined broadly or narrowly. This definition should not be made by employees, but instead through explicit institutional policies after consideration of environmental and occupational risks. Without such a policy (or with a poorly communicated one), employee practices may, in effect, unnecessarily broaden the definition of infectious waste, thereby increasing infectious waste volumes and costs.

VOLUME AND WEIGHT REDUCTION TECHNIQUES

The advantage of employing any one of these techniques depends on the cost basis for waste disposal services. In many cases, disposal fees are based entirely on the weight of the waste, thus making volume reduction techniques of little use.

Both incineration and steam sterilization reduce the volume of medical wastes. Incineration reduces waste to ash, achieving about 70–90% reduction in solid waste volume. Removal of air and contraction of heat-labile plastics present in medical wastes are responsible for a 10–30% volume reduction during steam sterilization.

Compaction, shredding, or grinding is the most commonly used pretreatment volume reduction technique. The compaction process, however, may present its own hazards (e.g., formation of aerosols) and usually hinders the efficiency of steam sterilization and incineration.

REFERENCES

1. U.S. Congress, Office of Technology Assessment. "Serious Reduction of Hazardous Waste" OTA-ITE-317 (Washington, DC: U.S. Government Printing Office, September 1986).
2. Rutala, W. "Cost-Effective Application of the Centers for Disease Control Guidelines for Handwashing and Hospital Environmental Control." *Am. J. Infectious Control* 13:218–224 (1984).
3. Lupin, R. and M. Sprafka. "A Waste Reduction Project Within the University of Minnesota Hospitals." University of Minnesota Physical Plant (June 10, 1980).
4. Marks, C. H. "Burn or Not to Burn: The Hospital's Modern-Day Dilemma." *Poll. Eng.* (November 1988), pp. 97–99.
5. U.S. Congress, Office of Technology Assessment. "Issues in Medical Waste Management—Background Paper." OTA-BP-O-49 (Washington, DC: U.S. Government Printing Office, October 1988), p. 3.

SECTION III

Management of Other Medical Waste

CHAPTER 11

Antineoplastic Drugs and Other Chemical Wastes

Hospitals and other medical institutions generate relatively small volumes of chemical waste. This does not mean that its disposal is a trivial matter; management of chemical waste in a manner that protects human health and the environment often requires a great deal of effort and incurs significant costs. Safe management of chemical wastes also requires prevention of harmful exposures to personnel. Readers should employ the chemical safety measures described in Chapter 14.

FEDERAL, STATE, AND LOCAL REGULATIONS

The explanation of the legal requirements for hazardous waste management in this chapter is not complete. Federal hazardous waste laws cover a wide variety of activities, but only those common to medical institutions are discussed here. State and local rules that differ from federal standards are also not discussed here. The regulation of chemical waste is extremely complicated. This chapter tries to aid the reader in understanding these regulations and serves as an introduction to reading the actual rules and understanding the regulatory details. The authors urge readers to contact their local, state, and federal hazardous waste agencies (see Appendix A) for guidance. Also enlist the assistance of the institutional legal counsel; keep him or her informed of chemical waste management practices.

RCRA and Superfund

EPA regulates today's hazardous waste practices under the authority of the Resource Conservation and Recovery Act and its amendments. Most of the following discussion of hazardous waste regulation is based on these laws. The Comprehensive Environmental Response, Compensation, and Liability Act (CERCLA, often referred to as Superfund), as amended by the Superfund Amendments and Reauthorization Act (SARA), regulates harm caused to the environment from hazardous waste, particularly from dumps, spills, and leaking landfills. As mentioned in Chapter 1, Superfund makes generators perpetually liable for their hazardous waste. This is of particular concern to medical institutions. As explained in this chapter, many medical institutions are subject to reduced requirements under RCRA. However, reduced regulation under RCRA does not reduce the institutional liability specified under Superfund. Moreover, compliance with past or present laws may have no bearing on liability if environmental harm can be proven to have resulted from an institution's wastes or practices. Chemical waste that is not regulated and practices allowed under RCRA can cause environmental harm; if they do, actions under Superfund can make the institution pay for cleanup. Also, medical institutions generate unique types of chemical waste that can be easily traced to the institution. Because of this, many hospitals, clinics, and laboratories take only partial advantage of the reduced requirements for small quantity generators, and instead manage their waste in ways that best limit their liability.

REGULATION OF CHEMICAL WASTE

The Environmental Protection Agency establishes the minimum standards for chemical waste management and disposal nationwide. States may adopt stricter standards. Therefore, *readers must contact their state hazardous waste authority* (*see Appendix A*) *to obtain the correct and complete understanding of their institution's requirements under the law.* EPA's regulations are explained here. Because most state rules are similar, the discussion of regulations in this chapter will provide readers with a good basis for understanding differences in their state's regulation of hazardous waste. In some states there is no substantive difference. A few states have no hazardous waste laws of their own; generators in these states are subject only to the EPA rules.

EPA regulates chemical waste if it has a hazardous characteristic or is listed in 40 CFR 261.* Lists are principally used to regulate toxic chemicals. (See Table 11.1.) Chemical waste that has one or more of the described characteristics or is on one of EPA's lists is legally identified as hazardous waste. (The term "hazardous waste" is frequently used generically for waste that has some

---

*40 CFR 260 is an abbreviation for Title 40 of the *Code of Federal Regulations*, Part 260. Although these regulations take considerable effort to read and understand, we urge the reader to obtain a copy, as well as a copy of the equivalent state regulations. The CFR is useful even where an equivalent state regulatory code exists, as state rules and disposal contractors may refer to it.

Table 11.1.   Examples of Chemical Waste Regulated by EPA[a]

Wastes Exhibiting These Characteristics	Examples
Ignitability	alcohol, acetone, xylene
Corrosivity	chromic acid, muriatic acid
Reactivity	Dry picric acid
Production of toxic leachate	barium sulfate, mercury

Listed Hazardous Wastes	Examples
Spent solvents (no longer usable)	toluene, chloroform
Discarded commercial chemical products	
Acute hazardous waste	carbon disulfide, cyanide salts
Toxic waste	acrylamide, ethylene oxide, formalin – see Table 11.3

[a]See 40 CFR 262.20 through 40 CFR 262.33(f) for the complete criteria and lists.

hazardous properties, an interpretation wider in scope than EPA's legal definition. In this chapter, "hazardous waste" refers only to waste meeting EPA's definition.) Note also that only waste is regulated in 40 CFR 260-270; unused materials and surplus drugs that may be used further, for example, are not wastes and therefore are not regulated. An expired drug that cannot be used or a chemically contaminated disposable item would be defined as waste.

Characteristic Hazardous Waste

Chemical waste is regulated as hazardous waste if it exhibits one of four characteristics: ignitability, corrosivity, reactivity, or ability to produce toxic leachate in a landfill (which may contaminate groundwater). EPA defines the criteria for these characteristics in Subpart C of 40 CFR Part 261. Few drugs exhibit any of these characteristics. Other chemicals that are found in medical institutions, however, do exhibit one of these characteristics: acids used for cleaning are corrosive; solvents used in laboratories are typically ignitable; expired mercury thermometers from nursing stations are toxic leachate-producing waste. Disposal of these chemicals must be done according to EPA rules as described below.

Listed Hazardous Waste

EPA also regulates some chemical wastes that are discarded commercial chemical products (DCCP) listed in 40 CFR 261.33(e) as acute hazardous wastes (having hazardous waste numbers beginning with P) or in 40 CFR 261.33(f) as toxic wastes (numbered with U). Discarded commercial chemical products are commercially pure chemicals, any technical grade of the chemical, or formulations in which the chemical is the sole active ingredient. An example of a pure grade DCCP is an unwanted drug in the manufacturer's source container. Because formaldehyde is listed as a toxic waste (U122), waste formalin (a 37% solution of formaldehyde used for disinfection) is a technical-grade DCCP. The rules also cover residue remaining in a non-empty container and debris resulting from the cleanup of a spill. The important point to note is that some wastes are not regulated. Wastes resulting from the use of these materials (products of a process or operation) are *not* regulated, such as materials contaminated with a listed chemical generated in the course of a standard procedure.

ANTINEOPLASTIC DRUGS AND OTHER CYTOTOXIC AGENTS

Among medical institutions, particular concern has focused on the handling and disposal of antineoplastic drugs and other cytotoxic agents used in medicine.[1] Most of these drugs can reasonably be expected to be mutagenic, teratogenic, and/or carcinogenic to both man and animals. It should also be noted that many drugs—not just those used for chemotherapy—are toxic, and their disposal deserves extra care. This is especially true for larger quantities of expired or surplus drugs.

Identification of Antineoplastic Drug Waste

Antineoplastic drug waste comes in many forms. Perhaps the wastes of most serious concern are unused portions of source containers (containers in which drugs are supplied), expired drugs, and surplus admixtures, which typically have larger quantities or higher concentrations of the drug. Chemically contaminated waste is also generated, including used needles and syringes, tubing and bottles used for intravenous administration, empty drug vials and ampoules, gloves, aprons, and disposable benchtop coverings from biological safety cabinets. Needles, and perhaps some other items, may be considered both chemically contaminated waste and infectious waste. Disposal of wastes with multiple hazards is the topic of Chapter 13.

Regulation of Antineoplastic Drug Wastes

No antineoplastic agents exhibit a characteristic of a hazardous waste, but seven commonly used antineoplastic drugs are listed by EPA as discarded commercial chemical products.[1] (See Table 11.2 and 40 CFR 261.33(f).) Each is listed as a toxic waste. (This list is also referred to as the "U" list.) As described above, EPA regulates these seven drugs only when they are waste commercial products (e.g., unemptied discarded source containers). Supplies contaminated with these drugs during preparation and administration are not regulated. Thus, the vast majority of wastes from the use of antineoplastic drugs and most cytotoxic agents used in medicine are not regulated by EPA. (These wastes *are* regulated under some state laws, however; check with your state hazardous waste authority.)

Table 11.2. Antineoplastic Drugs Listed by EPA as Toxic Wastes[a]

Drug	Hazardous Waste Number
Mitomycin C	U010
Chlorambucil	U035
Cyclophosphamide	U058
Daunomycin	U059
Malphalan	U150
Streptozotocin	U206
Uracil mustard	U237

[a]This pertains only to discarded commercial chemical products and spills resulting from these materials.

Lack of regulation does not mean that these wastes are without hazard. EPA's rules were written with the intent to primarily regulate industrial hazardous waste—the nation's most pressing waste management need. Nevertheless, improper disposal of unregulated chemical wastes—including antineoplastic drugs—can result in harm to human health and the environment and liability for an organization. One liability concern is that antineoplastic drugs and their associated waste (e.g., tubing, IV bottles, labeled containers) are readily traceable to medical facilities. Because of this, many medical institutions have chosen to manage all of their antineoplastic waste as regulated hazardous waste. Others have analyzed the hazard of their wastes and have selected disposal routes that minimize both threats to the environment and their liability. Selection of disposal routes is discussed below.

OTHER CHEMICAL WASTE

Medical institutions commonly generate other types of chemical wastes. Laboratories can be the source of a great variety of hazardous wastes (see Table 11.3), and they typically generate the largest volumes of hazardous waste within medical institutions. Maintenance activities also generate hazardous wastes, such as discarded cleaning compounds, degreasing solvents, paints, thinners, and water treatment chemicals.

Table 11.3. Hazardous Wastes Generated by Laboratories[a]

Waste	DOT Shipping Name	Hazard Class	UN/NA ID Number
Solvents			
Acetone	Waste Acetone	Flammable Liquid	UN1090
Benzene	Waste Benzene	Flammable Liquid	UN1114
Chloroform Trichloromethane	Waste Chloroform	ORM-A	UN1888
1,4-Dioxane Diethylene ether 1,4-Diethylene oxide Diethylene oxide Dioxyethylene ether	Waste Dioxane	Flammable Liquid	UN1165
Ethanol Ethyl alcohol Grain alcohol	Waste Ethyl Alcohol	Flammable Liquid	UN1170
Ethyl ether Ether Diethyl ether Diethyl oxide	Waste Ethyl Ether	Flammable Liquid	UN1155
Formalin Formaldehyde solution (1) flash point greater than 141°F	Waste Formaldehyde Solution	ORM-A (or Combustible Liquid if shipped in containers larger than 110 gallons)	UN2209
(2) flash point less than or equal to 141°F	Waste Formaldehyde Solution	ORM-A (or Combustible Liquid if shipped in containers larger than 110 gallons)	UN1198
Hexane n-Hexane	Waste Hexane	Flammable Liquid	UN1208
Isopropanol Isopropyl alcohol TPA Dimethyl carbinol 2-Propanol	Waste Isopropanol	Flammable Liquid	UN1219
Methanol Methyl alcohol Wood alcohol	Waste Methyl Alcohol	Flammable Liquid	UN1230
Methyl ethyl ketone MEK 2-Butanone	Waste Methyl Ethyl Ketone	Flammable Liquid	UN1193
Methylene chloride Dichloromethane	Waste Dichloromethane (or Waste Methylene Chloride)	ORM-A	UN1593

Table 11.3, continued

Waste	DOT Shipping Name	Hazard Class	UN/NA ID Number
Pentane	Waste Pentane	Flammable Liquid	UN1265
Petroleum ether	Waste Petroleum Ether	Flammable Liquid	UN1271
Tetrahydrofuran THF	Waste Tetrahydro-furan	Flammable Liquid	UN2056
Toluene Toluol Methyl benzene	Waste Toluene	Flammable Liquid	UN1294
Xylene Xylol Dimethyl benzene	Waste Xylene	Flammable Liquid	UN1307
Carbon tetrachloride Carbon tet Tetrachloromethane Perchloromethane	Waste Carbon Tetrachloride	ORM-A	UN1846
Ignitable liquids	Waste Flammable Liquids, NOS	Flammable Liquid[b]	UN1993
	Waste Combustible Liquids, NOS	Combustible Liquid[b]	NA1993
Acids/Bases			
Acetic acid	Waste Acetic Acid, Glacial	Corrosive Material	UN2789
	Waste Acetic Acid, Solution	Corrosive Material	UN2790
Hydrochloric acid	Waste Hydrochloric Acid	Corrosive Material	UN1789
Nitric acid	Waste Nitric Acid, over 40%	Oxidizer	UN2031
	Waste Nitric Acid, 40% or less	Corrosive Material	NA1760
	Waste Nitric Acid, Fuming	Oxidizer	UN2032
Perchloric acid	Waste Perchloric Acid, not over 50% acid	Oxidizer	UN1802
	Waste Perchloric Acid, exceeding 50% but not exceeding 72% acid	Oxidizer	UN1873
	Waste Perchloric Acid, exceeding 72% acid	Forbidden[c]	

Table 11.3, continued

Waste	DOT Shipping Name	Hazard Class	UN/NA ID Number
Sulfuric acid	Waste Sulfuric Acid	Corrosive Material	UN1830
	Waste Sulfuric Acid, Spent	Corrosive Material	UN1832
Oleum Fuming sulfuric acid	Waste Oleum	Corrosive Material	NA1831
Ammonium hydroxide Ammonia solution Aqueous ammonia	Waste Ammonium Hydroxide, con-taining less than 12% ammonia	ORM-A	NA2672
	Waste Ammonium Hydroxide, con-taining not less than 12% but not more than 44% ammonia	Corrosive Material	NA2672
Potassium hydroxide Caustic potash	Waste Potassium Hydroxide, Solid	Corrosive Material	UN1813
	Waste Potassium Hydroxide, Liquid	Corrosive Material	UN1814
Sodium hydroxide Caustic soda Lye	Waste Sodium Hydroxide, Solid	Corrosive Material	UN1823
	Waste Sodium Hydroxide, Liquid	Corrosive Material	UN1824
Non-Specific Wastes			
Corrosive liquids	Waste Corrosive Liquids, NOS	Corrosive Material	UN1760
Corrosive solids	Waste Corrosive Solid, NOS	Corrosive Material	UN1759
Oxidizer, corrosive, liquid	Waste Oxidizer, Cor-rosive, Liquid, NOS	Oxidizer	NA9193
Oxidizer, corrosive, solid	Waste Oxidizer, Cor-rosive, Solid, NOS	Oxidizer	NA9194
Oxidizer	Waste Oxidizer, NOS	Oxidizer	UN1479
Poisonous liquid[d]	Waste Poison B, Liquid, NOS	Poison B	UN2810
Poisonous solid	Waste Poison B, Solid, NOS	Poison B	UN2811
Corrosive, poisonous liquid	Waste Corrosive Liquid, Poisonous, NOS	Corrosive Material	UN2922

Table 11.3, continued

Waste	DOT Shipping Name	Hazard Class	UN/NA ID Number
Poisonous, corrosive solid	Waste Poisonous Solid, Corrosive, NOS	Poison B	UN2928
Poisonous, oxidizing liquid	Waste Oxidizer, Poisonous, Liquid, NOS	Oxidizer	NA9199
Poisonous, oxidizing solid	Waste Oxidizer, Poisonous, Solid, NOS	Oxidizer	NA9200
Hazardous waste	Hazardous Waste, Liquid, NOS	ORM-E	NA9189
Hazardous waste[e]	Hazardous Waste, Solid, NOS	ORM-E	NA9189

[a]These descriptions may change given variations in waste characteristics, conditions, or process modifications.

[b]Substances with a flash point less than 100°F are classified as "Flammable Liquid"; substances with a flash point greater than or equal to 100°F and less than 200°F are classified as "Combustible Liquid."

[c]Forbidden materials are prohibited from being offered or accepted for transportation.

[d]Certain gases and volatile liquids (e.g., cyanogen, phosgene) are classed as Poison A. If classed as Poison A, the gases and liquids have a different UN/NA ID: NA1953 for poisonous liquid or gas, flammable, NOS; or NA1955 for poisonous liquid or gas, NOS.

[e]Materials (e.g., disposable labware) contaminated with small quantities of a variety of hazardous substances can generally be classified as Hazardous Waste, NOS, unless a more specific DOT shipping name applies. The entire weight of the contaminated material, not just the weight of the substance(s) making it hazardous, is considered when determining quantity.

Regulation of Miscellaneous Chemical Wastes

Many laboratory and maintenance wastes are hazardous because they are ignitable or corrosive. Spent organic solvents (e.g., alcohols, acetone, petroleum distillates) are hazardous waste. Old picric acid from the microbiology laboratory that has been allowed to dry is potentially explosive (reactive characteristic). Chromic acid is sometimes used to clean laboratory glassware. Waste chromic acid is regulated as a hazardous waste because it is a dangerous oxidizer and contains chromium, thereby meeting the criteria of being corrosive, being reactive, and producing toxic leachate.

Laboratories also generate discarded commercial chemical products in the form of expired reagents. Other examples are given in Tables 11.1 and 11.3.

REGULATION OF INSTITUTIONS GENERATING HAZARDOUS WASTE

EPA regulates hazardous waste from cradle to grave. This means generators, transporters, and facilities that treat, store, and dispose of it are all regulated.* Figure 11.1 displays how wastes and generators are regulated. Generators are allowed to accumulate hazardous waste within certain limits (see Table 11.4) and perform hazard reduction procedures under certain conditions. Beyond those limits, accumulation becomes regulated as storage, which requires a permit. Facilities that treat, store, or dispose of hazardous waste (TSD facilities) are required to obtain a hazardous waste permit from EPA or a license from the state, which is a lengthy, arduous process. These permits are specific for hazardous

Table 11.4. Summary of EPA Generator Regulations[a]

Requirement	Fully Regulated Generator (>1000 kg/month)	Generator of 100–1000 kg/month	Conditionally Exempt Small Quantity Generator (<100 kg/month)	CFR Reference (Title 40)
Waste determination	Yes	Yes	Yes	262.11
Identification no.	Yes	Yes	No	262.12
Use manifests	Yes	Yes[c]	No	262.20
Permitted TSDs only	Yes[d]	Yes	No[e]	262.20(b)
Follow DOT[f] rules	Yes	Yes	No	262.30
Accumulation limit[g]	90 days or 1000 kg	180–270 days or 6000 kg	1000 kg	262.34(a), (d) and (e) 261.5(g)(2)
Mark containers	Yes	Yes	No	262.34(2) and (3)
Preparedness and prevention	Yes	Yes	No	262.34(a)(4) and (d)(4)
Contingency planning	Yes	Reduced	No	262.34
Training	Yes	Yes	No	262.34(a)(4)
Recordkeeping	Yes	Yes	No	262.40
Reporting	Yes	No	No	262.41

[a]State rules differ; please check.
[b]And generators of more than 1 kg/month of an acute hazardous waste.
[c]No if a waste is being sent offsite to be reclaimed under a routine contracted service.
[d]If allowed by state regulations, solid waste facilities may be used for acute hazardous waste generated by conditionally exempt small quantity generators.
[e]State-licensed solid waste facilities may be used.
[f]U.S. Department of Transportation rules for transport of hazardous materials (49 CFR 171–173).
[g]See Table 11.6.

---

*By EPA's definition, incineration is an example of a treatment method; placement of waste in a landfill is disposal.

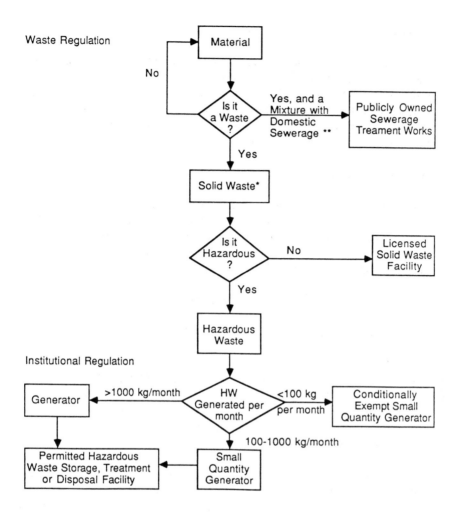

* Includes solids, liquids and gases; technical term for materials that are to be discarded, or have their intended purpose (such as garbage, refuse and normal trash).
** Rules and ordinances of the Publicly Owned Treatment Works must be followed.

Figure 11.1. Regulation of generators of chemical waste and hazardous waste.

waste and have little to do with permits for emission to the air or fire-protected storage of flammable materials.

Hazardous waste management activities at medical institutions are typically restricted to generation. It is extremely rare (and generally not advised by the authors) for a minor hazardous waste generator to obtain a permit for transport, treatment, storage, or disposal.

Hazardous Waste Determination

All institutions—no matter how small—are required to examine their wastes and determine if any are regulated hazardous waste. All solid wastes (normal trash, garbage, rubbish) should be scrutinized. The examination can most easily be done by examining materials being purchased and the hazard information contained on their material safety data sheets. In some cases testing may be necessary to make this determination. Tables 11.1–11.3 should also be helpful.

Generator Status

Generators of hazardous waste are regulated according to the amount of hazardous waste generated each month. Institutions generating less than 100 kg (220 lb, about one-half of a 55-gallon drum) of hazardous waste per month are conditionally exempt small quantity generators. Most hospitals and many other medical institutions fall into this regulatory category.

Conditionally Exempt Small Quantity Generators

In general, conditionally exempt small quantity generators must meet only two criteria:

1. Accumulated hazardous waste cannot exceed 1000 kg (2200 lb).
2. Waste must be disposed of in an EPA-permitted or state-licensed hazardous waste facility; a facility permitted, licensed, or registered by a state to manage municipal or industrial solid waste; or an approved recycling facility.

A solid waste facility is one that manages normal trash, such as a local sanitary landfill. Because of the reduced hazard of smaller quantities, conditionally exempt small quantity generators are allowed to dispose of their hazardous waste along with their normal trash in states where the regulations are no more stringent than EPA's, *if* the solid waste facility accepts hazardous waste from conditionally exempt small quantity generators, which is very rare. As explained previously, this practice carries liability for both the generator and the facility. As described below, there are other options, including commercial disposal services.

Small Quantity Generators

Generators of 100 to 1000 kg/month (220 to 2200 lb per month, or about one-half to five 55-gallon drums) must comply with the following requirements (which are listed in Table 11.5 with references to the CFR for a more detailed explanation):

- Waste containers must be managed safely (e.g., closed), marked as hazardous waste, and dated.
- Contingency preparations are required, but are less stringent than those for generators.
- All waste must be treated or disposed of at a facility specifically permitted or licensed for hazardous waste.
- More than 6000 kg of hazardous waste may not be accumulated at any time. Depending on the distance to the nearest TSD facility, the time limit for accumulation ranges from 180 to 270 days. (See Table 11.5.)
- U.S. Department of Transportation rules for marking, labeling, placarding, and packaging must be followed for shipping hazardous waste. Manifests must be used to document each shipment (see Figure 11.2) and kept for three years.

Thus, it is nearly impossible for generators of more than 100 kg/month to remain in compliance without the services of a hazardous waste disposal company. Contractual service for routine pickup is usually the best system. Many firms help their customers complete manifests and comply with DOT rules. Chapter 17 has suggestions for dealing with commercial disposal firms.

Table 11.5.   Accumulation Time Limits for Generators of 100 to 1000 kg/month[a]

Transport Distance	Accumulation Limit	CFR Reference
Less than 200 miles	180 Days	40 CFR 262.34(d)
More than 200 miles	270 Days	40 CFR 262.34(e)

[a]The quantity of waste accumulated onsite may never exceed 6000 pounds.

Acute Hazardous Waste

Acute hazardous waste is regulated separately in 40 CFR 261.5(e) because of its high hazard. Examples of these wastes that may be present at medical-related facilities include cyanide salts (used as analytical reagents), sodium azide formulations (used as a preservative), osmium tetroxide (used for electron microscopy), and carbon disulfide (a laboratory solvent). Generation of more than one kg per month of an acute hazardous waste brings a generator under the full regulation for those wastes, including requirements for:

- a 90-day limit on accumulation
- a written plan for contingencies
- training personnel to implement the contingency plan
- use of manifests for shipments

Please print or type   (Form designed for use on elite (12-pitch) typewriter.)   Form Approved OMB No 2050-0039 Expires 9-30-91

UNIFORM HAZARDOUS WASTE MANIFEST	1 Generator's US EPA ID No	Manifest Document No	2 Page 1 of	Information in the shaded areas is not required by Federal law

3. Generator's Name and Mailing Address

A. State Manifest Document Number

B. State Generator's ID

4. Generator's Phone (   )

5. Transporter 1 Company Name   6.   US EPA ID Number

C. State Transporter's ID

D. Transporter's Phone

7. Transporter 2 Company Name   8.   US EPA ID Number

E. State Transporter's ID

F. Transporter's Phone

9. Designated Facility Name and Site Address   10.   US EPA ID Number

G. State Facility's ID

H. Facility's Phone

11. US DOT Description (including Proper Shipping Name, Hazard Class, and ID Number)	12 Containers		13 Total Quantity	14 Unit Wt Vol	I. Waste No
	No	Type			

G E N E R A T O R

a.

b.

c.

d.

J. Additional Descriptions for Materials Listed Above

K. Handling Codes for Wastes Listed Above

15. Special Handling Instructions and Additional Information

16. GENERATOR'S CERTIFICATION: I hereby declare that the contents of this consignment are fully and accurately described above by proper shipping name and are classified, packed, marked, and labeled, and are in all respects in proper condition for transport by highway according to applicable international and national government regulations

If I am a large quantity generator, I certify that I have a program in place to reduce the volume and toxicity of waste generated to the degree I have determined to be economically practicable and that I have selected the practicable method of treatment, storage, or disposal currently available to me which minimizes the present and future threat to human health and the environment, OR, if I am a small quantity generator, I have made a good faith effort to minimize my waste generation and select the best waste management method that is available to me and that I can afford.

Printed/Typed Name	Signature	Month Day Year

T R A N S P O R T E R

17. Transporter 1 Acknowledgement of Receipt of Materials

Printed/Typed Name	Signature	Month Day Year

18. Transporter 2 Acknowledgement of Receipt of Materials

Printed/Typed Name	Signature	Month Day Year

F A C I L I T Y

19. Discrepancy Indication Space

20. Facility Owner or Operator Certification of receipt of hazardous materials covered by this manifest except as noted in Item 19

Printed/Typed Name	Signature	Month Day Year

EPA Form 8700-22 (Rev 9-88) Previous editions are obsolete

Figure 11.2. Hazardous waste manifest.

Note that it is possible for a generator to be classified as a conditionally exempt small quantity generator but still be under full regulation for its acute hazardous waste.

Requirements for generators of more than 1000 kg/month (full regulation) are outlined and referenced in Table 11.4.

WASTE MINIMIZATION

The practice of preventing hazardous waste generation is called waste minimization. Examples include:

- recycling and reuse
- substitution of a hazardous material with a nonhazardous one
- recovery of heat from waste solvents or silver from photography labs
- redistribution of surplus materials
- restricting purchases to allow orders for only what is immediately needed

Waste minimization is by far the best method of managing hazardous waste. It is wise to review waste minimization efforts periodically. Compare efforts with those of similar institutions generating similar wastes. Much progress is being made in this area; firms are offering a wider array of recycling services and substitute products are being developed.[2] The following are descriptions of waste minimization practices that have reduced hazardous waste generation at medical institutions.

Minimizing surplus materials: Surplus chemicals are common in laboratories. Larger quantities than necessary are purchased for cost savings, then become waste when they get old (i.e., degraded chemicals that are unusable) or when procedures change. When disposal costs are included, "large economy size" containers of chemical nearly always cost more.[3] Instead of allowing these surplus materials to become wastes, find another laboratory that can use the chemicals and escape regulation, liability, and disposal costs.

Solvent recycling/heat recovery: Stills can be purchased that safely distill waste xylene (and other solvents) to sufficient purity for reuse; this is especially useful for histopathology laboratories. Be sure that the still has automatic emergency shutoff in case of overheating. Commercial distillation services are also available, as are companies that use waste solvents as a fuel.

Product substitution: Material substitution is increasingly available. Less hazardous cleaning compounds and formulations have been developed that successfully replace chromic acid and degreasing solvents.

Silver recovery: This can be accomplished in photography laboratories (including X-ray facilities) by the purchase or rental of a recovery unit. Some silver recovery companies pay institutions for the installation and operation of recovery units.

DISPOSAL OPTIONS FOR CHEMICAL WASTES

Generators have many choices of methods to treat or dispose of their hazardous waste. Management activities can take place onsite (at the institution) or offsite (at a commercial facility). In most cases, a variety of treatment and disposal techniques are employed, depending on what is most suitable for the waste type and the institution.

As discussed in Chapter 1, many factors should be considered when choosing a disposal route. Onsite methods, for instance, provide greater institutional control. Landfilling has been a frequently used offsite disposal method, but it is the disposal route of the greatest liability. EPA is concerned about the potential of landfills to contaminate groundwater, and consequently has banned many wastes from landfills. Waste mercury solutions are generated in some lab procedures, and the banning of this waste from landfills has caused difficulty for some generators. Onsite chemical treatment procedures, referenced below, offer an alternative.

The following is a description of chemical disposal options available to most medical institutions, in approximate order of increasing difficulty. Although it may be obvious, note that in all cases heat, steam, and gas sterilization can never be used for the disposal of chemical wastes. It is also forbidden for a generator to avoid regulation by mixing or diluting regulated hazardous waste with a nonregulated waste, even if the mixture does not have a hazardous characteristic.

Disposal to the Normal Trash

If state and local laws allow, unregulated waste and (on rare occasions) hazardous waste from conditionally exempt small quantity generators can be disposed of in a licensed solid waste facility (i.e., a facility that handles normal trash, such as the local sanitary landfill). When considering this option institutions must first understand the liability risks of this practice. Some hospitals restrict normal trash disposal to unregulated wastes of very low hazard, such as disposable gloves, benchtop coverings, or IV tubing having a low concentration of a toxic chemical. Some solid waste haulers and landfill operators may not allow use of their facilities for any chemical wastes, thus preventing an institution from taking advantage of these reduced requirements.

Sanitary Sewer

In most municipalities, disposal of chemicals into the sanitary sewer is regulated by the sanitary sewerage district or a similar agency operating the local publicly owned treatment works (POTW) that services the institution. By exempting wastewater mixtures from the definition of solid and hazardous waste, EPA allows disposal of waste chemicals into the sanitary sewer as long as they comply with the POTW's limits and restrictions. This allowance is found at 40 CFR 261.4:

§261.4 Exclusions.

(a) *Materials which are not solid wastes.* The following materials are not solid wastes for the purposes of this part:

(1)(i) Domestic sewage; and

(ii) Any mixture of domestic sewage and other wastes that passes through a sewer system to a publicly-owned treatment works for treatment. "Domestic sewage" means untreated sanitary wastes that pass through a sewer system.

Some POTWs allow very little or no use of the sanitary sewer system for chemical disposal. Use of less efficient systems and systems that mix storm and sanitary sewer effluents is severely restricted. POTW rules that allow waste disposal take advantage of the system's ability to degrade organic chemicals (by microbiological action) and to dilute toxic chemicals to insignificant concentrations. Most cytotoxic agents are quite reactive and will readily break down in an efficient POTW. When allowed, disposal to the sanitary sewer is only acceptable for dilute solutions of chemicals that are easily degraded or of low toxicity.

The above discussion applies only to wastewater effluents to the local POTW. Other discharges of wastewater effluent from the institution, such as a storm sewer, must not be used for chemical disposal.

Medical-Pathological Incineration

Many medical institutions have access to a medical or pathological incinerator. These units are frequently located onsite and are used primarily for the disposal of pathological and infectious waste. The state or local environmental agency sometimes requires licensing or permitting these units for air emissions, solid waste disposal, or infectious waste management.

These incinerators differ greatly from incinerators that are designed and licensed specifically to burn chemical wastes. EPA issues permits for hazardous waste incinerators under extremely strict criteria, making it nearly impossible for all but large hazardous waste disposal firms or chemical companies to site, permit, and operate a hazardous waste incinerator. Only a few medical institutions in the United States have an EPA permit or equivalent state license to incinerate hazardous waste. Those that do are limited to a few wastes and/or are associated with large organizations. Medical or pathological incinerators, which are not designed to meet such stringent standards, must not be used for the disposal of regulated chemical waste; without the proper permit, it would be illegal.

However, some medical institutions employ medical-pathological incinerators for unregulated wastes containing organic chemicals in small amounts and low concentrations. In contrast to disposal in a landfill or sanitary sewer, incineration destroys chemicals, although destruction efficiency is likely to be less than at an EPA-permitted hazardous waste incinerator.

Return to Manufacturer

Hospitals have reported success in returning expired/surplus drugs to the manufacturer for disposal.[4] Also, some gas manufacturers will take back unemptied cylinders.

Chemical Treatment

Chemical treatment of hazardous waste is a regulated activity, requiring either an EPA permit or state license. Medical laboratories, however, can take advantage of two exceptions allowed by EPA and most state agencies.

First, since elementary neutralization is not considered a treatment process (see 40 CFR 264.1(g)(6)), waste mineral acids such as sulfuric, nitric, and hydrochloric acid can be neutralized to solutions of mineral salts and disposed of in the sanitary sewer. Bases like ammonium hydroxide can also be neutralized.

Second, totally enclosed treatment facilities are exempt from permitting requirements. (See 40 CFR 264.1(g)(5).) According to *Prudent Practices for the Disposal of Waste Chemicals from Laboratories,* there is a common understanding that "laboratory procedures for the reduction or destruction of the hazardous characteristics of a chemical is part of the experiment and does not constitute 'treatment' in the regulatory sense."[5] Thus, colleges, universities, and clinical laboratories have argued that laboratory procedures that include treatment methods to manage wastes qualify for this exception. In addition, in-lab chemical treatment is not considered treatment per se, but rather a hazard reduction procedure, consistent with EPA's intent to minimize as much waste as possible.

Chemical treatment methods are perhaps the most promising disposal option for antineoplastic drugs and cytotoxic agents. These methods, which have been developed and verified by several research groups in recent years,[6-8] are especially useful for wastes that are expensive and difficult to dispose of otherwise. In-lab treatment methods have also been published for a variety of other drugs and chemicals common to research and teaching laboratories.[5,8,9] Although they may initially seem imposing to the pharmacist and clinical laboratory technician, most of the methods have been designed to be performed safely by a science professional.

Commercial Chemical Waste Incinerator

Two dozen or so commercial hazardous waste incinerators are operating nationally. These facilities represent a disposal route of versatility, very low liability, high cost, and increasing popularity. Even toxic metal waste can be incinerated;

the metals concentrate in the ash for further treatment and disposal. Some facilities recover heat from the waste. EPA's stringent permitting and operating regulations assure a 99.99% destruction efficiency for organic chemicals. Since most generators have few or no other commercial chemical treatment options, incineration is the method of choice for offsite disposal.

Commercial disposal firms are available to transport, treat, and dispose of hazardous waste. Sometimes these firms share facilities, so a generator's waste may be handled by several different companies. It has been only in recent years that most firms have turned their attention to servicing smaller generators of hazardous waste. Some are doing an excellent job of servicing specialized wastes, such as recyclable solvents. Medical facilities in metropolitan areas have a variety of companies and disposal routes to choose from, and price competition lowers their costs. In rural areas, the cost of shipment makes offsite hazardous waste disposal an extra burden. Companies typically charge a flat pickup fee, as much as $1000. Commercial disposal services are disproportionately expensive for nearly all small quantity generators. These facts make onsite disposal options extremely valuable for smaller generators. Chapter 17 discusses reducing offsite commercial disposal costs by forming cooperative agreements and joint contracts with other small quantity generators.

Difficult Wastes

Potentially explosive, reactive, and unknown wastes are difficult to dispose of, although commercial disposal firms are increasingly able to manage these wastes. Chemical treatment methods have been suggested for some reactive chemicals,[5,9] but the handling and disposal of explosives is best left to the local bomb squad. (Regulatory approvals may have to be obtained.) A procedure for identification of unknowns (e.g., substances whose containers are old or poorly labeled) is given in *Prudent Practices*.[5]

GETTING HELP

Advice is an important resource when attempting to cope with these complex regulations and hazardous materials. Start with the institution's legal counsel or risk manager. Discussions with safety officers from other hospitals, colleges and universities, and trade organizations allow comparisons of successes and failures. Inspectors from the state hazardous waste agency are valuable sources of information; they should have a list of available transporters, recyclers, and disposal firms. Ask for a copy of their inspection form so you can perform an unofficial compliance audit. EPA operates a useful toll-free hotline (1-800-424-9346) and has available many useful publications for small quantity generators.[10]

Chapter 17 advises on the best ways of working with commercial disposal firms.

REFERENCES

1. Vaccari, P. L., K. Tonat, R. DeChristoforo et al. "Disposal of Antineoplastic Wastes at the National Institutes of Health." *Am. J. Hosp. Pharm.* 41:87–93 (1984).
2. U.S. Environmental Protection Agency, Risk Reduction Engineering Laboratory, Office of Research and Development. "Guide to Waste Minimization in Selected Hospital Waste Streams" and "Guide to Waste Minimization in Research and Educational Institutions." U.S. EPA, Washington, DC (1990).
3. American Chemical Society. "Less Is Better" (Washington, DC: ACS, 1985).
4. Gregoire, R. E., R. Segal, and K. M. Hale. "Handling Antineoplastic-Drug Admixtures at Cancer Centers: Practices and Pharmacist Attitudes." *Am. J. Hosp. Pharm.* 44:1090–1095 (1987).
5. National Research Council. *Prudent Practices for Disposal of Waste Chemicals from Laboratories* (Washington, DC: National Academy Press, 1983).
6. Armour, M. A., R. A. Bacovsky, L. M. Browne, P. A. McKenzie, and D. M. Renecker. *Potentially Carcinogenic Chemicals Information and Disposal Guide* (Edmonton, Alberta, Canada: University of Alberta, 1986).
7. Lunn, G. and E. B. Sansone. "Reductive Destruction of Dacabazine, Procarbazine Hydrochloride, Isoniazid and Iproniazid." *Am. J. Hosp. Pharm.* 44:2519–2524 (1987).
8. International Agency for Research on Cancer. "Laboratory Decontamination and Destruction of Carcinogens in Laboratory Wastes: No. 37 (Aflatoxin B1, B2, G1, G2), No. 43 (N-Nitrosamides), No. 49 (Polycyclic Aromatic Hydrocarbons), No. 54 (Hydrazines), No. 55 (N-Nitrosamides), No. 61 (Haloethers), No. 64 (Aromatic Amines and 4-Nitro Biphenyl), No. 73 (Antineoplastic Agents)" (Lyon, France: IARC).
9. Armour, M. A., L. M. Browne, and G. L. Weir. *Hazardous Chemicals Information and Disposal Guide,* 2nd ed. (Edmonton, Alberta, Canada: University of Alberta, 1984).
10. Office of Solid Waste and Emergency Response. "Understanding the Small Quantity Generator Hazardous Waste Rules: A Handbook for Small Business." EPA/530-SW-86-019 U.S. EPA, Washington, DC (1986).

CHAPTER 12

Managing Low-Level Radioactive Waste

Radioactive materials have proven to be valuable tools in medicine. At hospitals and clinics, radionuclides make the diagnosis of disease easier and have made new treatments possible. At school and research laboratories, radionuclides are used to measure minute amounts of biochemicals and act as metabolic tracers, thereby broadening our understanding of life and disease. All of these practices generate low-level radioactive waste.

Radioactive waste need not be a serious problem for medical institutions. Planning can reduce costs and the amount of waste generated.[1-3] When implemented, a comprehensive waste management plan will entail separating different types of radioactive waste and using several management pathways, including waste minimization and a diversity of treatment and disposal methods. Administrative support, adequate space, and the cooperation of institutional waste generators, waste handlers, and safety staff are necessary.

Two difficulties overshadow plans for radioactive waste management. First, radioactive waste management is perhaps the nation's most politically sensitive waste problem. People are generally both fearful of and uneducated about radiation. They frequently confuse low-level radioactive waste issues with those connected with high-level radioactive waste and nuclear power. This field also has its own jargon, making it all the more confusing.

Second, the rules for low-level radioactive waste management are inconsistent. Federal, state, and municipal laws all differ from each other; regulations and regulatory authority differ depending on the use and identity of radioactive materials. Institutional licenses for the use of radioactive materials have conditions for waste management that vary greatly—even for similar institutions in the same city.

Because of these variations, this chapter is limited to a general description of low-level radioactive waste management practices and federal requirements. Solutions are suggested, but the details vary for each institution and the specific rules that apply to it. All radioactive material users are subject to the waste management methods specified by federal, state, and local laws, as well as the conditions specified by their licenses.

LOW-LEVEL RADIOACTIVE WASTE

Low-level radioactive waste is a broad and somewhat ambiguous category, defined not by what it is, but what it is not. It includes the least harmful types of radioactive waste. Low-level radioactive waste is defined in the Federal Low-Level Radioactive Waste Policy Act of 1980 as radioactive waste not classified as high-level, transuranic, spent fuel, or certain by-product material.[4] Excluded waste types are within the domain of nuclear power plants, fuel facilities, etc. (although these facilities also generate low-level radioactive waste). Most generators of infectious and medical waste generate only low-level radioactive waste.

In the context of all radioactive waste generated, low-level radioactive waste is characterized by high volume and very low hazard. Through 1983, low-level radioactive waste made up 84% of the total volume of all radioactive wastes.[5] By volume nuclear power plants are the largest source of low-level radioactive waste. Of the low-level radioactive waste disposed of by shallow land burial, medical and educational institutions accounted for only 14%.[6]

The degree of hazard posed by different types of radioactive waste can be crudely compared according to their activity as measured in becquerels (Bq). Through 1983 low-level radioactive waste made up only 0.1% of the total activity of all radioactive wastes.[5] The hazard of radioactive waste additionally depends upon the type of radionuclide in the waste. Because low-level radioactive waste is primarily composed of radionuclides with low radiotoxicity (e.g., H-3), there is a very large difference in the potential hazards presented by high-level and low-level radioactive waste.

For the purposes of regulating waste management practices at commercial waste management sites, 10 CFR 61 divides low-level radioactive waste into Classes A, B, and C (Class A being least dangerous). These classifications must be used when describing and labeling waste for offsite shipments.

Composition of Low-Level Radioactive Waste

Despite its exclusions, the definition of low-level radioactive waste includes a great variety of waste. Radioactive waste from medical institutions is comprised of any waste containing or contaminated with radioactive isotopes (radionuclides). Medical institutions use radionuclides in therapy, diagnosis (e.g., nuclear medicine, radioimmunoassays), and research.[1-3] Table 12.1[1,2,7,8] lists the radionuclides

Table 12.1.   Radionuclides Commonly Used in Laboratories and Medical Facilities[1,2,7,8]

Radionuclide	Half-Life	Uses[a]
H-3	12.3 y	N, L, R
C-14	5730 y	R
P-32	14 d	R
S-35	87 d	R
Ca-45	163 d	R
Cr-51	28 d	R
Co-57	270 d	L
Co-60	5.3 y	N, S
Ga-67	3.3 d	N
Tc-99m	6 hr	N
Mo-99	2.8 d	N
In-111	2.8 d	N, S
I-123	13 hr	N
I-125	60 d	N, L, R, S
I-131	8 d	N
Xe-133	5.3 d	N
Cs-137	30 y	N, S
Y-169	32 d	N
Ir-192	74 d	S
Th-201	3 d	N

[a]L = Clinical laboratory.
N = Nuclear medicine.
R = Research.
S = Used as a sealed source.

used in medical institutions and their uses. Note that the majority of radionuclides used in medicine have half-lives of 60 days or less. (All radioactive isotopes [radionuclides] decay over time to a different isotope that may or may not be radioactive. The time it takes for half of the radionuclides to decay is known as its half-life. Seven half-lives reduce activity to 1% of its original level and ten to 0.1%.)

Low-level radioactive waste includes the following waste types:[6]

- *Dry waste*: This includes disposable gloves, aprons, and benchtop covers; emptied containers; and other paper, glass, and plastic disposable items contaminated with radionuclides.[7] (Some sources use the term "solid" waste for dry waste. This may be confusing because EPA defines "solid waste" as a liquid, gas, or solid waste.) Low-level radioactive wastes shipped to disposal sites are comprised of about 50% dry waste by volume.[2,7]

- *Liquid scintillation cocktail and vials*: Liquid scintillation cocktail (LSC) is used to measure very low levels of radioactivity. The sample to be measured and cocktail are mixed in a vial and placed in a liquid scintillation counter. Many research projects and diagnostic tests depend on the use of radiolabeled chemicals and this measurement method. The vials are glass or plastic. LSC typically contains H-3 or C-14.

- *Other liquids:* Aqueous wastes containing radioactive materials are generated in research and diagnostic laboratories. To a lesser extent, research generates some organic solvent-based liquid waste that is not liquid scintillation fluid.
- *Animal carcasses and tissues*: Sometimes called biological waste, animal carcasses and tissues containing radioactive materials are generated from research and teaching functions at medical institutions.
- *Sealed sources and targets*: These wastes, which tend to be of fairly high activity relative to other types of low-level radioactive waste, are generated in low volume from larger medical and research laboratories.

REGULATION OF LOW-LEVEL RADIOACTIVE WASTE

Rules for low-level radioactive waste management may originate in federal or state regulations, local ordinances, the institution's license conditions, or (when waste is shipped offsite) requirements of the waste management contractor. In addition, radioactive waste (radwaste) regulation and management depends upon its identity, origin, amount, and concentration.

The U.S. Nuclear Regulatory Commission (NRC) has the primary authority among federal agencies for setting standards for the safe use of radioactive materials in the United States. General NRC rules for waste management are in 20 CFR parts 301 through 311. EPA presently has a less extensive role in regulating environmental impacts of radioactive materials, although some recent proposed rules would restrict the release of waste gases and incinerator emissions.[9]

State Regulatory Agencies; Agreement States

Any state may establish its own program to regulate and license users of radioactive materials. As of June 1987, 29 states have done so. These states are called ''agreement states.'' State regulations must not be less stringent than the federal rules. Most state rules are identical or very similar to NRC rules. However, there are many instances where state and local governments (in both agreement and non-agreement states) have set standards for low-level radioactive waste that are more strict than the NRC. For example, New York City did not accept the NRC rule change that allowed certain types of liquid scintillation cocktail to be disposed of without regard to radioactivity.[9] Institutions that use radioactive materials must ascertain if there are state or local regulations that apply to low-level radioactive waste management.

Types of NRC Licenses

Although the NRC does not require licensing under some circumstances (i.e., when *only* exempt quantities and devices are used), most medical institutions using radioactive materials will need to obtain an NRC or a state license. Diagnostic kits need to be licensed under 10 CFR 31.11 but are subject to reduced requirements. Some vendors of radioactive materials are required by their licenses to have proof of the institution's radioactive material license prior to shipment.

There are several types of byproduct material (radionuclides produced by reactors) licenses issued by the NRC that may apply to generators of infectious and medical waste:

- *General Licenses* are issued to physicians, veterinarians, clinical laboratories, and hospitals for use of small quantities of radioactive material in in vitro clinical or laboratory tests not involving administration to humans. General licenses issued under 10 CFR 31.11 are the least strict with regard to disposal of radioactive waste.
- *Specific Medical Use Licenses* are issued to physicians in private practice as well as medical institutions. Medical use pertains to the intentional internal or external administration of radioactive materials or radiation to humans.
- *Broad Scope Licenses* are another type of specific license issued to institutions that provide patient care and conduct research using radionuclides.

For large institutions engaging in many operations that use radionuclides, a single license may include a variety of uses.

Exempt, Deregulated, or *De Minimis* Wastes

Certain wastes are exempt from regulation. Institutions that use radionuclides for in vitro clinical or laboratory testing or possess less than 200 μCi (and stay within specific radionuclide limits) may dispose of all of their low-level radioactive waste in the normal trash (10 CFR 31.11). This allows General Licensees with I-125 to dispose of contaminated waste from radioimmunoassay kits as normal trash. Some medical institutions plan their laboratory operations around these limits to take advantage of this exemption. As explained in detail below, liquid scintillation cocktail and animal carcasses contaminated with H-3 or C-14 below 1.85 kBq/g have been deregulated. Many experts have suggested that the NRC expand this exemption to other types of waste and additional radionuclides.[1,7,10]

If adopted by the NRC and states, these wastes would be designated as *de minimis* (trivial) or below regulatory concern (BRC). Such a ruling would also relieve the stress on regional waste management facilities.

Lack of regulation doesn't mean lack of work; waste generators must continue to document compliance and show that the wastes meet the exemption criteria.

Low-Level Radioactive Waste Policy Act

In 1980, Congress passed the Low-Level Radioactive Waste Policy Act (amended in 1985), which requires each state to be responsible for the management of its waste.[4] The act empowers the states of Washington, Nevada, and South Carolina (the only existing disposal sites) to exclude out-of-state waste after January 1, 1993. The law encourages the formation of multistate compacts to cooperatively manage low-level radioactive waste and develop regional waste management sites. (Some compacts and their sites are not really "regional"—California and North Dakota are in the same compact!)

The law requires states and their compacts to meet certain milestones in developing waste management plans and regional sites. Financial penalties are assessed on the states and their waste generators for not meeting these milestones and, in some cases, the existing sites can refuse access before the 1993 deadline.

It is unlikely that many sites will be in operation by 1993, potentially leaving most waste generators without access to disposal sites. The final deadline for operation of the new low-level radioactive waste management facilities is January 1, 1996. Afterwards, ownership of and liability for the waste is transferred to the state in which it was generated.

Performance-Based Standards

Some parts of the NRC regulations specify detailed procedures, while other parts state limits or standards that must be met and leave the procedural details to the licensee. In the license application, the institution proposes its intended procedures and disposal methods to meet the performance standards. Approval of the license is contingent on NRC's approval of the application's proposals. License conditions are often negotiable. The NRC has published a series of Regulatory Guides and model procedures for certain performance-based standards, including a model procedure for waste management.[11] Thus, in many cases institutions can choose one of several waste management options—or invent a new method—as long as it meets the NRC performance standards and is approved.

The ultimate test of all radiation safety programs is the NRC philosophy that licensees should go beyond the regulatory standards to "make every reasonable effort to maintain radiation exposures, and releases of radioactive materials in effluents to unrestricted areas, as low as is reasonably achievable." The key word here is "reasonable."[10] Is a "reasonable exposure" zero or inconsequential? At what cost? These questions must be addressed when planning for radioactive waste management, but there are no definitive answers.

The Mixed Waste Problem

Some radioactive wastes have chemical characteristics that call for additional treatment and disposal precautions. For example, toluene-based liquid scintillation cocktail is radioactive, toxic, and ignitable. Waste that is regulated due to its radioactivity and chemical hazard is called "mixed waste," which is discussed in detail in Chapter 13. Another type of multiple-hazard waste discussed in Chapter 13 is radioactively contaminated infectious waste.

PLANNING FOR LOW-LEVEL RADIOACTIVE WASTE MANAGEMENT

Low-level radioactive waste management at larger medical institutions typically combines several onsite methods with offsite disposal. As discussed in Chapter 1, such diversity will help the plan to be flexible. Planning will require preparing an institutional waste profile and describing the types and volumes of waste generated. Consult similar institutions; decisions are difficult to make and similar institutions are likely to adopt similar policies.

Source Separation

Source separation is a prerequisite for an efficient waste management plan. Whenever possible, keep infectious waste, chemical waste, and normal trash separate from radioactive waste. Further, the NRC recommends that radioactive waste should be separated by radionuclide, level of activity, and waste type (e.g., aqueous vs organic solvents) as much as possible.[11] Source separation enables an institution to find the easiest and least costly management method for each waste type.

Space and Time Requirements

Waste management requires secure and separate work areas for temporary storage, sorting, handling, and packaging waste for shipment, so that unauthorized personnel cannot easily gain access. As explained below, additional space is required to hold waste for decay. As with all aspects of radiation safety, waste management entails a great deal of paperwork to track waste and document compliance. Nearly every step of waste handling requires checking containers with a survey meter. Liquids may need to be sampled and counted. Institutional administrators must recognize and budget for the labor and space requirements of waste management.

Regulatory Labyrinth

For some waste management methods (particularly incineration) regulatory approvals may be difficult to obtain. Cooperation between federal and state agencies responsible for radioactive materials and environmental protection is poor. Some state agencies aggressively intervene in radioactive waste management, while others are reluctant to review and approve. Approvals are sometimes contingent on the approval of another agency, which can put the generator in a Catch-22 predicament.[2,10] For example, an NRC license to incinerate radioactive waste may be conditional on approval by the state air quality agency, which in turn may wait for the NRC's final license issuance.

Strategy for Low-Level Radioactive Waste Management

Three strategies for treating or disposing of low-level radioactive waste are common:[1,5]

- *Storage for decay* to deregulated levels: The NRC calls this Decay in Storage (DIS).
- *Dilution and dispersal:* Permitted releases to the environment include disposal of aqueous liquids to the sanitary sewer, evaporation of aqueous waste to the atmosphere, and incineration. Radioactivity cannot be destroyed or reduced by any process or treatment method. Rather, radioactive waste that is treated (e.g., by incineration) or disposed of (e.g., in the sanitary sewer) is diluted (by combustion gases and other wastewater, respectively) and dispersed into the environment. (Incineration also leaves radionuclides in the ash.) Regulators rely on dilution and dispersal to keep exposures to the public as low as is reasonably achievable. Dilution and dispersal is most appropriate for wastes that can be easily diluted to background levels. (Radiation is natural and pervasive in our environment: this is called background radiation. Dilution and decay can reduce the level of radiation in waste to levels that are indistinguishable from background levels.)
- *Concentration and containment:* Accomplished by transferring the waste to a burial site or a regional waste management facility. Because radionuclides decay to nonradioactive daughters, radioactivity is reduced when waste is stored or buried. For low-level radioactive waste of relatively high activity and hazard (e.g., sealed sources and surplus source materials) the optimal management method is to keep the waste separate and concentrated, then ship the waste to a regional disposal facility.

Overall, the NRC instructs licensees that "in all cases, consider the entire impact of various available disposal routes. Consider occupational and public exposure to radiation, other hazards associated with the material and routes of disposal (e.g., toxicity, carcinogenicity, pathogenicity, flammability), and expense."[11]

The following sections explain general alternatives of onsite radioactive waste management. Waste management practices for specific waste are then covered: liquid scintillation cocktail, animal carcasses and tissue, and aqueous liquids.

ONSITE WASTE MANAGEMENT ALTERNATIVES

Because radioactive waste generated by medical facilities is characterized by short half-lives, low activity, or low radiotoxicity, most of it can be managed and disposed of onsite. All medical generators should have a place to hold short–half-life waste for decay in storage. Further, some radioactive wastes generated by medical institutions (e.g., high-activity flammable solvents) have no existing commercial disposal option and no such capabilities are being planned by compacts. Thus, for the foreseeable future, each institution is expected to completely manage these wastes itself.

Decay in Storage

Radioactive materials with short half-lives should be held in storage onsite until the activity of the waste is reduced to background levels. This is a common practice for the many wastes generated from medical diagnosis and therapy that have half-lives of 60 days or less (see Table 12.1); waste is held for decay to background levels and then disposed of as normal trash. Decay in storage is becoming the accepted management method for P-32 (14-day half-life), Cr-51 (28-day) and I-125 (60-day) wastes, and in some cases even S-35 (87-day) and Ca-45 (163-day) wastes. One medical center anticipated a 50% reduction in the amount of radioactive dry waste needing shipment to a commercial disposal facility as a result of decay in storage.[2] License requirements for storing low-level radioactive waste for decay are often similar to the requirements of 10 CFR 35.92 (for specific medical use licenses), which allow decay in storage and disposal in the normal trash for wastes containing radionuclides with half-lives of less than 65 days as long as:

- The waste is held for at least 10 half-lives. (In 10 half-lives, activity is reduced to 0.1%.)
- The waste is monitored before disposal to the normal trash, and its radioactivity cannot be distinguished from the background radiation level.
- All radiation labels are removed or obliterated.

Taking advantage of this provision requires recordkeeping, manpower, and a special area for decay in storage of radioactive waste. Areas used by institutions for decay in storage range in design from a single room within a hospital to a separate, remote building. Space requirements can be extensive.[3] All facilities need to be secure from access by unauthorized personnel and must provide some means of limiting exposures to the nearest point of human occupancy (either by shielding or distance). If radioiodine is to be stored there (e.g., I-125), sufficient ventilation, a well-dispersed exhaust, and monitoring may be necessary.

Incredibly, there are many instances of institutions shipping short–half-life wastes to a low-level radioactive waste landfill because storage space for decay has not been made available.[2,10] The unwillingness of an institutional administrator to provide space for this simple and cost-effective alternative indicates environmental irresponsibility. The current crisis in low-level radioactive waste stems, in part, from institutions that have shirked the responsibility of onsite waste management. Nobel Laureate Rosalyn S. Yalow, who has been outspoken on the need for rational radwaste management, has opined that "certainly it should be possible for all institutions to arrange for onsite decay of radionuclides with half-time of 2–3 months or less" and that "there should be relatively little radioactive waste from biomedical institutions."[10]

Sanitary Sewer

Most institutions can and do use the sanitary sewer for disposal of aqueous wastes. Keep in mind:

- Release to the sanitary sewer is regulated according to 10 CFR 20.303 and, in some instances, by state rules and institutional license conditions.
- Local sanitary sewer ordinances may further restrict the release of radionuclides.
- These rules rely on the dilution that occurs in the sewer system, so that radioactivity is undetectable when it reaches the wastewater treatment plant. For dilution to occur, wastes disposed of in the sanitary sewerage must be readily soluble or dispersable in water. If this does not occur, radionuclides can reconcentrate in sewerage sludge.[12]
- In addition to activity limits of individual radionuclides, there are daily and monthly activity and concentration limits based on the total sanitary sewerage release of the institution. See 10 CFR 20.303.
- All sanitary sewers are not alike. Some municipal systems mix storm and sanitary wastewater and may bypass wastewater treatment facilities in the event of heavy precipitation.
- Sanitary sewer ordinances may also regulate the liquids that contain the radioactive materials. For example, formalin (a mixture of formaldehyde and water) may be disposed of in the sanitary sewer in some municipalities, but not in others. Also, sewer ordinances limit the release of certain solutes. Although many chemicals safely degrade in the sanitary

sewer, these limits must be checked before disposing of wastes into the sanitary sewer system. See Chapter 11 for further discussion of disposal of chemical wastes into the sanitary sewer.

• Much of the radioactivity administered to patients ends up in the sanitary sewer. This is a standard practice that is exempt from regulation. For most medical institutions, releases to the sanitary sewer from administration to patients is far greater than from other waste disposal.

Evaporation

Evaporation is a potential disposal route for small amounts of aqueous waste.[11] This is usually carried out by storing an open container of liquid waste in a fume hood or other ventilation device. Exhaust to an unrestricted area is limited to the permissible concentration limits in Table II of Appendix B to 10 CFR 20. Limits can be met by restricting amounts or by engineering controls. A license condition may require routine monitoring of these releases to the atmosphere. Also, state or local air quality rules may restrict this practice, especially with respect to evaporation of wastes that contain volatile organic chemicals.

Incineration and Fuel Use

Few generators have access to an incinerator specifically designed for radioactive waste. Instead, a small number of medical institutions have an onsite incinerator designed for some waste types that are common in the radioactive waste stream. A common example is a pathological incinerator that can be used for animal carcasses and tissue contaminated with radionuclides.[6]

Incineration has the advantage of reducing volumes—up to 98% in one case.[3] Institutions have also been reported to use incineration for dry wastes instead of direct disposal to the normal trash to prevent misunderstandings with local landfill operators.[3] Incineration of radioactive waste is not specifically authorized in the *Code of Federal Regulations*. Instead, 10 CFR 20.305 requires that an institution apply separately for permission to incinerate waste, giving special attention to compliance with respect to releases to unrestricted areas. Stack gas releases to the environment and exposures to the public are typically estimated by calculation using conservative assumptions and modeling[13] or by actual measurement.[14] Onsite incineration may be restricted by specific license conditions to certain radionuclides and instantaneous, daily, monthly, and/or annual activity and concentration limits. Radionuclides vary in volatility, which results in some percentage of total activity remaining in the ash. Disposition of the ash (i.e., normal trash or offsite disposal) depends upon the resulting radionuclide concentration and its solubility.[15] As before, ash that contains short–half-life radionuclides should be held for decay.

Pathological incinerators are typically of a starved-air design and are not necessarily appropriate for combustion of other waste types. For example, paper and plastic dry waste are difficult to burn and may produce excessive particulate and acidic emissions. Chapter 7 discusses these limits and incinerator operation in detail.

Another thermal treatment option for low-level radioactive wastes composed of flammable organic solvents (such as liquid scintillation cocktail) is to use them as a fuel, either by blending the wastes with fuel oil or directly injecting them into a boiler. Federal rules and many states allow this practice (with certain limitations and required approvals) as long as its purpose is to recover energy from the waste. However, some states require the boiler to meet chemical waste management standards if the waste has the characteristics of a regulated chemical waste.

Disposal to the Normal Trash

The following types of waste contaminated with radionuclides can be disposed of in the normal trash if allowed by license conditions and state and local laws:

- exempt waste, such as in vitro kits generally licensed under 10 CFR 31.11
- waste that has decayed 10 half-lives and is indistinguishable from background levels
- deregulated waste, such as liquid scintillation counting vials that have been emptied of fluid
- other low-concentration dry wastes (as allowed by license conditions) that meet NRC release limits for unrestricted areas

The NRC calls disposal in the normal trash "release to in-house waste."[11] Liquids should not be disposed of in the normal trash. Aqueous liquids in vials and tubes should first be decanted into the sanitary sewer.

Waste Minimization Techniques

Substitution, source separation, and procedural changes have all been shown to be effective ways of minimizing radioactive waste. Also, compaction has been shown to reduce offsite disposal costs.[2] Substituting a less hazardous product for a more hazardous one has minimized liquid scintillation counting waste (see below). Nonradioactive diagnostic laboratory tests have been introduced that can replace procedures that formerly required radionuclides.

Perhaps most importantly, keep nonradioactive trash out of the radioactive waste stream. The NRC urges licensees to "remind employees that nonradioactive waste such as leftover reagents, boxes, and packing materials should not be mixed with radioactive waste."[11] Changes in procedures and methods can reduce waste volumes. Counting methods are available that do not use liquid scintillation cocktail (especially for gamma emitters[2]) or that employ micro-samples that reduce the volume and cost of waste disposal more than 95%.[16] Regulatory Guide 10.8 recommends that institutions "occasionally monitor all procedures to ensure that radioactive waste is not created unnecessarily. Review all new procedures to ensure that waste is handled in a manner consistent with established procedures."[11]

Indefinite Storage

It is unlikely that many state or compact waste management facilities will be operational before the three existing disposal sites become inaccessible in 1993. If timetables proceed as scheduled, generators of low-level radioactive waste will be required to store waste onsite until new waste management facilities are operational. A complicating factor is that medical and research institutions tend to generate unusual wastes. These wastes are characterized by the presence of uncommon radionuclides or additional hazards (e.g., chemicals or infectious agents). For example, liquid radioactive wastes that have been deregulated and cannot be disposed of in the sanitary sewer (because of their organic nature), have historically been disposed of by absorption and burial.[6] The compacts may not provide any offsite disposal option for liquids. Most plans for regional low-level radioactive waste management sites don't account for these unusual wastes. This will leave their generators with no choice but to manage them onsite, typically by means of indefinite storage.

OFFSITE WASTE MANAGEMENT ALTERNATIVES

Medical institutions rely on offsite disposal for those wastes that cannot be managed onsite. Because of the radionuclides used, research wastes tend to have longer half-lives than other medical wastes; much low-level radioactive waste shipped offsite originates from research.[2] Dry waste that cannot be held for decay is sent to a radioactive waste burial site. Animal carcasses and tissue are sometimes packed in lime and buried.[6]

Shallow Land Burial

Historically, shallow land burial has been the principal means of offsite disposal for low-level radioactive waste[5] and continues to be an important waste disposal method. Disposal entails placing waste containers in excavated trenches (called "cells") below soil grade. When filled with waste, the cells are capped by soil. To prevent liquids from leaching into groundwater beneath the site, liquid wastes are now restricted and must be absorbed or otherwise solidified, but many older sites are contaminated. Since 1980 only three commercial shallow land burial sites have been available to generators.

Because of its potential for environmental impact and long-term liability, shallow land burial should be considered the waste management method of last resort.

It is only appropriate for wastes that cannot be managed onsite or sent to an off-site treatment facility (e.g., incinerator)—and waste burial may not be available after 1993.

Alternative Management Methods

Above-ground containment (e.g., vaults) and facilities that allow retrievable storage are being considered as an alternative to shallow land burial by most compacts. Such a facility would be easier to monitor and less likely to impact ground-water and surface water and would permit the retrieval of wastes that have decayed to background levels. Incineration has great potential for treating and reducing the volume of radioactive waste. To date, however, commercial incineration is only available for liquid scintillation cocktail (see below) and few compacts are considering a regional incinerator.

Commercial Waste Management Firms

A number of commercial waste management firms offer offsite storage, treatment, and disposal services for low-level radioactive waste. These firms operate as waste brokers by collecting waste from many small generators and contracting with treatment and disposal sites. Brokers must be licensed by the NRC (the NRC sometimes calls them "transfer agents") and may offer waste packaging as well as transport. Some brokers only handle prepackaged waste, while others may store, treat, or repackage wastes. It is cost-effective to consolidate wastes from various sources; administration is more efficient, and shipping costs to remote sites are less.

It is the responsibility of the waste generator to comply with the U.S. Department of Transportation rules for preparing waste for shipment (i.e., packaging, labeling and marking, etc.) However, most brokers provide these services. The broker will likely help prepare the manifests, which are shipping papers that describe and track the waste shipment. (See 20 CFR 311 and Figure 11.2.)

Chapter 17 discusses the use of commercial waste management firms and treatment and disposal facilities in more detail.

SOLUTIONS FOR SPECIFIC WASTES

Liquid Scintillation Cocktails and Vials

The high cost and difficulty of liquid scintillation cocktail (LSC) disposal has troubled many medical and academic generators. Before the mid-1980s, cocktails were largely composed of the organic solvents toluene, xylene, or, to a lesser extent, pseudocumene.

Waste management practices for LSC have evolved considerably in recent years. At first, many vials were shipped to a radioactive waste site for shallow land burial. The realization that burial of liquid organic waste threatened groundwater has made this practice undesirable. In addition, transportation of LSC waste to one of the few sites available for disposal is costly. With state and local approvals, many institutions use an onsite incinerator or boiler for treating liquid scintillation cocktail.[2] The vials are either emptied manually or by crushing for injection into the combustion chamber, or the entire case or tray of vials is batch-loaded into the incinerator. The advantages and disadvantages of incinerating radioactive waste are described above.

After 1980, EPA or state chemical waste regulations began to restrict or prohibit onsite fuel use and incineration of LSC. Liquid scintillation cocktail burned in a pathological incinerator does not meet EPA's criteria for chemical waste incineration. The NRC then promulgated the rule in 10 CFR 20.306 that allows the disposal of LSC below 1.85 kBq/g H-3 or C-14 without regard to radioactivity. This provision deregulated most LSC because it typically has extremely low levels of radioactivity. (Some states and locales continue to regulate these wastes as low-level radioactive waste.) However, toluene- and xylene-based cocktail, deregulated by the NRC, is regulated as chemical hazardous waste. Commercial incineration is an available, albeit expensive, method for treating this waste. See Chapter 13 for further discussion of mixed waste.

Recently, several manufacturers have introduced biodegradable, water-miscible cocktails that can safely be disposed of by discharge to the sanitary sewer. Tests show that in most instances, these cocktails perform as well or nearly as well as the older toluene-based liquid scintillation cocktails.[17] Sanitary sewer use still requires emptying vials by crushing, pouring, etc. Due to the limitations of most onsite incinerators for controlling emissions and efficient destruction, the authors prefer the use of biodegradable cocktail to onsite incineration.

Animal Carcasses and Tissue

If allowed by state and local laws, 10 CFR 20.306 allows the incineration of animal carcasses and tissue containing < 1.85 kBq/g H-3 or C-14 to be disposed of without regard to radioactivity. Animal carcasses are typically disposed of in an onsite incinerator. (See Chapter 7.) Institutions that do not have an onsite incinerator ship deregulated animal tissue to a regional commercial incinerator, grind the tissue for sewer disposal (see Chapter 8), or may continue to dispose of these carcasses in a radioactive waste landfill.

Higher-activity and other radionuclides can be incinerated onsite only with special permission (see above). Alternatively, animal tissue contaminated with short-half-life materials can be held for decay if frozen or refrigerated storage space is available.[3]

Aqueous Liquids

Liquid wastes made up of aqueous solutions of radionuclides and other solutes are prime candidates for sanitary sewer disposal as discussed above. Compliance with the daily and monthly limits may require the control of centralized collection and disposal. A program of source separation (such as providing laboratories with separate containers for aqueous radwaste and for organic solvent radwaste) may facilitate use of this route. To decrease releases, some institutions take the additional step of storing aqueous waste for decay before sanitary sewer disposal.

COMMENT ON THE PROBLEM

The controversy surrounding low-level radioactive waste is frustrating to many science and medical professionals. It is probably true that more is known about the hazards of radioactive materials than other hazardous materials such as asbestos, toxic chemicals, or infectious agents. Radiation can be quantified extremely accurately. Epidemiologists have extensively studied populations exposed to radiation and have been able to quite accurately establish the relationship of dose and incidence of cancer. This relationship becomes less clear near background levels of radiation and doses due to chronic exposures. However, the environmental and health risks of low-level radioactive waste can be measured more accurately than most other risks.

Dr. Yalow contends that "the problem of disposal of radioactive biomedical waste should be a nonproblem." The risks posed by the majority of low-level radioactive wastes appear to be insignificant compared to background sources of radiation and given what we know about low-level exposures.[10] Yet, progress in agreeing to define a *de minimis* or deregulated level and siting waste management facilities has been extremely slow. Even with such a detailed measurement of risk, the public does not rank the risks of low-level radioactive waste comparably with other risks of the same magnitude. Also, society is far from agreement as to what an acceptable risk is, even though the importance and need for radionuclides in medicine is well accepted. Instead, the issue of low-level radioactive waste has become an example of public communication that has somehow gone awry. Surely, we must find another process for making public policy if we are to succeed in solving similarly complex environmental problems.

REFERENCES

1. Wilkerson, A., R. C. Klein, E. Party, and E. L. Gershey. "Low-Level Radioactive Waste From U.S. Biomedical and Academic Institutions: Policies, Strategies, and Solutions," *Ann. Rev. Pub. Health* 10:299–317 (1989).

 2. Ovadia, J., and K. J. Francis. "Radioactive Waste Disposal in a Private Urban Academic Medical Center," in *Proceedings of Waste Management '83 Symposia on Waste Management* (Tucson, AZ: American Nuclear Society, 1983), pp. 201–203.
 3. Barish, E. L., J. L. Gilchrist, H. W. Berk, and R. O. Allen. "Radioactive Waste Management at Biomedical and Academic Institutions," paper presented at the International Conference on Radioactive Waste Management, Seattle, WA, May 16–20, 1983.
 4. Low Level Radioactive Waste Policy Act of 1980, Public Law 96-573.94.
 5. Carter, M. W., and D. C. Stone. "Quantities and Sources of Radioactive Waste," in *Proceedings of the 21st Annual Meeting of the National Council on Radiation Protection and Measurements* (Bethesda, MD: NCRP Publications, 1986), pp. 5–30.
 6. Vance, J. N. "Processing of Low-Level Wastes," in *Proceedings of the 21st Annual Meeting of the National Council on Radiation Protection and Measurements* (Bethesda, MD: NCRP Publications, 1986), pp. 38–53.
 7. Evdokimoff, V. "Dose Assessment from Incinerator of Deregulated Solid Biomedical Radwaste," *Health Physics* 52(3):325-329 (1987).
 8. Vetter, R. J. "Low-level Radioactive Waste: A National Disposal Problem," *Health and Environment Digest* 1(7):1–3 (1987).
 9. "National Emission Standards for Hazardous Air Pollutants; Regulation of Radionuclides; Proposed Rule," *Federal Register,* Vol. 54, No. 43 (March 7, 1989) pp. 9612–9668.
10. Yalow, R. S. "Disposal of Low-Level Radioactive Wastes: Perspective of the Biomedical Community," in *Proceedings of the 21st Annual Meeting of the National Council on Radiation Protection and Measurements* (Bethesda, MD: NCRP Publications, 1986), pp. 59–64.
11. "Guide for the Preparation of Applications for Medical Use Programs," NRC Regulatory Guide 10.8, Revision 2 (August 1987).
12. "Reconcentration of Radionuclides Involving Discharges into Sanitary Sewage System Permitted Under 10 CFR 20.303," NRC IE Information Notice No. 84-94, Revision 2 (December 21, 1984).
13. Philip, P. C., S. Jayaraman, and J. Pfister. "Environmental Impact of Incineration of Low-level Radioactive Wastes Generated by a Large Teaching Medical Institution," *Health Physics* 46(5):1123–1126 (1984).
14. Hamrick, P. E., S. J. Knapp, M. G. Parker, and J. E. Watson. "Incineration and Monitoring of Low-Level H-3 and C-14 Wastes at a Biological Research Institution," *Health Physics* 51(4):469–478 (1986).
15. Classic, K., G. Gross, and R. J. Vetter. "Solubility of Radionuclides in Ash from the Incineration of Animals," *Health Physics* 49(6):1270–1271 (1985).
16. Warner, G. T., and C. G. Potter. "New Liquid Scintillation Counter Design Erases Vial Disposal Problems," *Health Physics* 51(3):385 (1986).
17. Spate, V. L., and S. M. Langhorst. "A Comparison of the Counting Characteristics of Opti-Fluor® and Aquasol-2®: Liquid Scintillation Cocktail," *Health Physics* 51(5):667–671 (1986).

CHAPTER 13

Wastes with Multiple Hazards

Wastes with multiple hazards have more than one hazard—for example, they may be hazardous* and infectious, radioactive and infectious, hazardous and radioactive, or radioactive and hazardous and infectious. These wastes may have had multiple hazards at the time they were generated, or they may be mixtures of different wastes. From the perspective of sound waste management practice, the different waste streams should be kept separate; there is seldom a sound rationale for combining them.

Wastes with multiple hazards are sometimes referred to as "mixed wastes." This term can be confusing because it has been used for and now has the connotation of wastes with a radioactive component, specifically wastes that are both radioactive and hazardous.*

Wastes with multiple hazards often present special problems in management and disposal because of the difficulty in finding handling and treatment techniques that are compatible with all the hazards. Another source of difficulty may be the applicable regulations, because the regulatory requirements for the different waste streams may vary and may even be contradictory.

SOURCES OF WASTES WITH MULTIPLE HAZARDS

Medical wastes with multiple hazards are generated routinely during medical care. They are also generated during industrial and research activities. Most

---

*"Hazardous" means a regulated chemical waste.

multihazardous medical wastes have an infectious component. The most common type is probably the waste that is both infectious and radioactive; it is usually generated during diagnostic procedures or during patient care.

MANAGEMENT OF MULTIHAZARDOUS WASTES

Certain principles can serve as a guide for the management of wastes with multiple hazards. These are:

1. Comply with all relevant regulations.
2. Give priority to the hazard that presents the greatest risk.
3. Select treatment/management procedures that are compatible with *all* the hazards present in the waste.
4. If possible, select a treatment technique that will provide suitable treatment for all the hazards.
5. If necessary, provide additional treatment for eliminating the remaining hazards.

In addition, sound management strategies can aid in minimizing generation of multihazardous wastes and the problems associated with their disposal.

REGULATORY COMPLIANCE

The disposal of mixed low-level radioactive and chemical wastes poses a dilemma because the regulatory requirements of the U.S. Environmental Protection Agency for the disposal of hazardous chemical wastes and of the U.S. Nuclear Regulatory Commission for the disposal of radioactive wastes have been different and often conflicting. These two agencies are attempting to resolve the dilemma of conflicting jurisdictions so that generators will know how to dispose of commercial mixed wastes. See, for example, the 1987 joint policy statement on the disposal of commercial mixed low-level radioactive and hazardous wastes.[1]

The medical waste tracking regulations[2] address the issue of jurisdiction over regulated medical wastes that are also hazardous wastes.* In the covered states, all such "mixtures" (as they are referred to in the MWTA regulations) must be managed as hazardous wastes (i.e., per the RCRA hazardous waste regulations) if they are subject to the hazardous waste manifest requirements. Wastes that are exempt from the hazardous waste manifest requirements must be managed as regulated medical wastes, where applicable.

---

*See Reference 2, Section 259.31 ("Mixtures"), p. 12374.

Some hazardous waste treatment, storage, and disposal facilities do not accept mixtures and mixed wastes. Therefore, long-term storage may be necessary until a treatment/disposal option becomes available.

ASSIGNING PRIORITY TO GREATEST RISK

When wastes have multiple hazards, each waste must be evaluated individually to determine the proper management scheme for that waste. The best approach to the management of such wastes consists of four steps:

1. Ascertain which hazards are present in each waste.
2. Assess the relative degree of risk present in each hazard.
3. Assign priority to the hazard with the greatest risk.
4. Develop a management scheme based on the relative degrees of risk.

When a waste is radioactive, the radioactivity is usually considered the hazard of greatest concern. When these wastes have a relatively short half-life, the best approach is containment and storage of the waste until the radioactivity has decayed to background levels (see Chapter 12). At that time, the other hazard(s) can be addressed. Storage conditions must also be appropriate for containing/minimizing the other hazards.

Such a management approach is not necessarily the best. For example, if the waste is explosive or reactive, such a characteristic might take priority. With such wastes, detonation with containment and decontamination may be the best option.

COMPATIBILITY OF MANAGEMENT STRATEGY WITH ALL HAZARDS

The management techniques selected, whatever they may be, must be compatible with all the hazards present in the particular waste. The treatment for one hazard must not result in release of and exposure to any other hazard. Therefore, the selected treatment method must either treat all the hazards simultaneously or treat one hazard while having no adverse effect relative to the other hazard(s).

For example, steam sterilization might be suitable for treatment of the infectiousness of the waste, but this treatment method might have adverse effects if radionuclides or hazardous chemicals are also present. Steam sterilization must not be used if it will result in contamination of the sterilizer or volatilization of hazardous chemicals—events that could result in release and exposure.

Similarly, chemical treatment can have hazardous side effects. For example, use of hypochlorite to disinfect an aqueous solution can release radiolabeled iodine as a gas.

TREATMENT OPTIONS

The ideal treatment option is, of course, the one that simultaneously provides appropriate treatment to all the hazards. Proper incineration can be such a treatment option, because it is appropriate for infectious wastes, certain radionuclides, and certain solvents. For example, NRC regulations allow the incineration of some radionuclides within certain constraints. Similarly, a limited burn permit allows incineration of solvents that are hazardous only because of their flammability. (See Chapters 11 and 12 for details on incineration of chemical and radioactive wastes, respectively.) Any incineration of multihazardous waste (as well as chemical and radioactive wastes) must be within the limits of and in conformity with state permitting requirements.

Under some circumstances, no such "universal option" is available. For these multihazardous wastes, it may be necessary to provide sequential treatment steps that address the various hazards one at a time.

An example of this approach is the management of a waste that is both radioactive and infectious. The waste could first be stored for decay of the radionuclide (see Chapter 12), then steam sterilized to provide treatment for the infectiousness (see Chapter 6). For a waste that is both radioactive and hazardous, appropriate treatment might be storage for decay of the radionuclide followed by treatment and disposal in accordance with RCRA regulations for management of hazardous wastes (see Chapter 11).

MANAGEMENT STRATEGIES

The difficulties inherent in managing wastes with multiple hazards make it important to optimize waste management by minimizing the production of multihazardous wastes. Certain policies and activities will help to achieve this:

1. Do not mix waste streams.
2. Promote substitution policies.
3. Reduce the quantities of multihazardous wastes generated.
4. Identify sources of multihazardous wastes.
5. Store wastes for decay of radioactivity or until disposal options become available.

Prevention of the mixing of wastes is an important step in minimizing the quantity of multihazardous waste generated. Mixing can occur in two ways: (1) by the mixing of different waste streams (i.e., infectious wastes, hazardous chemicals, and/or radioactive wastes), thereby creating a multihazardous waste and (2) by the mixing of a multihazardous waste with any other waste, thereby increasing the quantity of multihazardous waste. Mixing of wastes should be prevented by establishing policy that prohibits such activities. Encourage implementation of this policy by providing special containers for each type of waste, including the multihazardous waste, that is generated in a particular area; the availability of special containers will help to prevent the mixing of wastes.

Substitution of materials is a method for eliminating the generation of some multihazardous wastes and for shifting the composition of others to wastes that are more easily managed. For example, nonhazardous chemicals should be substituted for hazardous chemicals whenever possible, thereby eliminating this component from the multihazardous waste. As another example, the substitution of a short–half-life radionuclide for one with a longer half-life will result in generation of a multihazardous waste that can be stored for a relatively short period until the radionuclide has decayed to background levels of radioactivity and the waste is no longer radioactive (and therefore not multihazardous).

Reduction in the quantities of multihazardous wastes generated is also important in optimizing management of these wastes. Review protocols to ascertain that they minimize the quantities of multihazardous wastes generated. Use procurement strategies to control quantities of materials ordered that lead to the generation of multihazardous wastes.

A review of activities at your institution or facility will help to identify sources of multihazardous waste. Once the sources are identified, the management strategies detailed above—that is, prevention of mixing of wastes, substitution of materials, and reduction of quantities of multihazardous wastes generated—can be implemented and supervised for these sources.

Long-term storage of multihazardous wastes, possibly onsite, may be an essential element in the management strategy for these wastes. Such storage may be necessary for two reasons: (1) to allow for decay of the radionuclides to background levels of radioactivity, and (2) to hold the wastes until commercial disposal alternatives are developed and become available. Any storage of multihazardous wastes must be under conditions that are compatible with the materials that will be held in storage. In addition, the storage area must be secure to prevent unauthorized access to the stored wastes.

REFERENCES

1. U.S. Environmental Protection Agency and U.S. Nuclear Regulatory Commission. "Guidance on the Definition and Identification of Commercial Mixed Low-Level Radioactive and Hazardous Waste," U.S. EPA OSWER Directive #9432.00-2, January 5, 1987. See also *Federal Register* 52(66):11147–11148, April 7, 1987.
2. U.S. Environmental Protection Agency. "Standards for the Tracking and Management of Medical Wastes; Interim Final Rule and Request for Comments," *Federal Register* 54 (56):12325-12394, March 24, 1989.

SECTION IV

Keeping Your System Going

CHAPTER 14

Occupational Safety for Waste Management

There have been references throughout the book to the risks involved in managing infectious and medical wastes. These are primarily occupational safety and health concerns because the risks are mostly those encountered by the waste handlers in the course of their work. It is important to recognize these risks and to include in the waste management plan practices that will improve occupational safety by reducing the risks.

In this chapter, the following topics are discussed:

- the occupational safety and health concerns associated with the management of infectious and medical wastes
- personnel who are at risk
- prudent practices for risk reduction
- other management techniques for risk reduction

OCCUPATIONAL SAFETY AND HEALTH CONCERNS

Two types of occupational safety and health concerns are associated with medical waste management: the risk of disease and the risk of injury.

The Risk of Disease

The risk of disease from infectious and medical wastes derives from the potential presence of three types of agents that can cause disease. These are:

- infectious agents
- toxic and hazardous chemicals
- radioactivity

The particular disease is specific to the causative agent. The risks of contracting a disease from waste handling are related to the nature of the causative agent present in the waste, the type and degree of exposure, and the health of the host.

Infectious Diseases

Transmission of infectious diseases is discussed in detail in Chapter 3. For the purposes of this chapter, it is sufficient to restate here that there are four possible routes of disease transmission:

- through the skin
- through mucous membranes
- by inhalation
- by ingestion

Each of these routes is a potential portal of entry through which infectious agents in waste can enter the body to cause disease in susceptible persons.

Two blood-borne infectious diseases are of particular concern today—hepatitis B and acquired immune deficiency syndrome (AIDS). Both are caused by blood-borne pathogens: by the hepatitis B virus (HBV) and the human immunodeficiency virus (HIV), respectively. Concern about these two diseases is sufficiently great that the U.S. Department of Labor (DOL) and the U.S. Department of Health and Human Services (DHHS) issued a Joint Advisory Notice on prevention of occupational exposures to these two viruses.[1,2] DOL's Occupational Safety and Health Administration is developing regulations that will require implementation of the recommendations.*

The statistics for illness among health-care workers support this concern. The Centers for Disease Control estimate that as many as 18,000 health-care workers per year may be infected by HBV, and that nearly 10% of these become long-term carriers of the virus.[2] Several hundred become acutely ill or jaundiced,[2] and 500–600 are hospitalized annually.[1] As many as 300 may die annually as a result of hepatitis B infections or complications[2] (from fulminant hepatitis, cirrhosis, and liver cancer).[1]

Occupational HIV infection has also been documented.[1,3,4] In these cases, infection resulted from contaminated blood splashed or rubbed onto broken skin or mucous membranes and from percutaneous transmission of the virus through needle sticks.

---

*At the time of book publication, these OSHA regulations were still in the proposed stage, and final regulations had not been promulgated. For the proposed rule, see Reference 3.

These are data for ''health-care workers''; separate data for handlers of infectious waste are not yet available.* It is prudent to assume that waste handlers are at risk for the same diseases as health-care workers when the infectious agents for these diseases are present in the waste.†

The infectious agents for these blood-borne diseases are transmitted by blood. The risk for transmission is greatest with contaminated sharps because these wastes can break the skin (through cuts, scrapes, and needle sticks) and convey the viruses and other infectious agents through the broken skin barrier.

Other infectious diseases can also be transmitted by wastes. Specimen cultures are of particular concern because they may have large numbers of infectious agents at high concentrations. Dusts, aerosols, and wet wastes also place waste handlers at risk for exposure to infectious agents.

The risk of exposure to infectious agents can be minimized by establishing and implementing a good infectious waste management plan. The plan should incorporate engineering controls, use of personal protective equipment, and appropriate procedures that will minimize the risks of occupational exposure. These would be required by the OSHA proposed rule.[3]

Exposure to Toxic and Hazardous Chemicals

Just as workers using hazardous chemicals are at risk of exposure, so are those who handle the waste chemicals. Exposures can be chronic or acute. Unsatisfactory waste management, including improper procedures, container use, and storage conditions, can lead to chronic exposures. Acute exposures usually result from the occurrence of a particular incident (such as a spill or a fire).

The type of illness caused by occupational exposure to toxic or hazardous chemicals depends on the particular chemical to which the worker is exposed and on the level of exposure.

The risk of exposure to toxic and hazardous chemicals can be minimized by the use of proper procedures for the handling, movement, and storage of waste chemicals. Emergency preparedness is also essential to ensure prompt and appropriate response to chemical spills. (See Chapter 15.)

Exposure to Radioactivity

Radioisotopes are commonly used at medical facilities in various diagnostic and treatment procedures. As a result, radioactive wastes are generated. Many facilities then store these wastes for prescribed periods of time to allow for decay of the radioactivity to acceptable levels before disposal.

---

*Occupational illness among waste handlers has not been studied. However, the Medical Waste Tracking Act of 1988 directs the Agency for Toxic Substances and Disease Registry (ATSDR) to study occupational disease among handlers of infectious waste.

†There are reports of general infection and hepatitis among waste handlers. These cases are almost impossible to document because they are usually settled out of court.

Waste handlers are at risk for exposure to radioactivity when the wastes are not managed properly. Examples of improper management that can result in exposure include:

- sloppy procedures that contaminate the external surface of the container
- use of improper storage containers that do not provide sufficient protection against exposure to radioactivity
- faulty records on waste generation dates and storage times that result in premature release of wastes from storage

The type of illness resulting from exposure to radioactivity is determined by the amount and type of exposure. Personal dosimetry badges are essential for monitoring the level of exposure for each worker who handles radioactive waste. Monitoring of personnel for exposure to radiation is required by NRC regulations[5] and by JCAHO standards.[6,7]

Exposure to radioactivity can be minimized by establishing and implementing a plan for the management of these wastes that provides appropriate procedures and safeguards.

The Risk of Injury

In addition to the risk of disease, workers are also at risk of injury from the handling of infectious and medical wastes.

Injury from Lifting and Handling Waste Containers

Back injuries and other muscle strains and sprains constitute a risk to workers who lift and handle waste containers. Two factors are common causes of such injuries. One is the container that is too large or too heavy to be lifted easily. The other is improper body movements and lifting techniques.

The risks of injury from the lifting and moving of waste containers can be minimized by the use of appropriate containers, by staff education that stresses the importance of using the designated containers for wastes, and by training waste handlers in the proper techniques for lifting and moving heavy containers.

Injury from Accidents

Another major cause of injury to waste handlers is accidents. There is an infinite list of potential accidents, but it includes such things as slips and falls, dropped containers, malfunctioning collection carts that cause containers to fall, and waste spills. The types of injury that result from accidents include muscle strains and sprains and broken bones.

Measures to prevent accidents include use of:

- containers that are appropriate for each kind of waste
- collection carts that are suitable for the types of waste containers used
- collection carts that are easy to load, move, unload, and clean
- adequate waste storage areas
- spill containment materials that are readily available

Injury from Sharps

A special hazard in infectious wastes is injury from sharps—the needles, scalpel blades, lances, and other such items used in medical care. Sharps probably constitute the greatest occupational hazard to health-care providers and waste handlers because of the double risk of injury and disease transmission. Sharps are discussed in detail in Chapter 3 under "Contaminated Sharps."

The risk of injury from sharps can be reduced by use of sharps containers to hold all waste sharps. The importance of using sharps containers must be stressed because such use offers protection to the health-care worker as well as to everyone who subsequently handles the waste. The medical waste tracking regulations require the use of sharps containers for sharps generated in the covered states,[8] and the proposed OSHA rule would require the use of sharps containers wherever contaminated sharps are generated.[3]

PERSONNEL AT RISK

It is essential that everyone realize that each individual's actions relative to waste management affect everyone else who subsequently handles that waste.

Health-Care Providers

Health-care providers—including nurses, physicians, medical technologists, and laboratory technicians—are the persons who generate the infectious and medical wastes. They are the ones who discard the materials that thereby become wastes.

Health-care providers are most at risk for exposure to wastes before they are placed into waste containers. Because of this, it is essential that wastes be discarded promptly, directly into the designated containers.

Housekeeping Personnel

Housekeeping personnel are at risk for occupational exposures to the hazards in infectious and medical wastes because they are the persons who handle the waste containers, collecting them from place of use and moving them through the facility to storage areas or treatment equipment. The principal hazard for these workers is waste that was not placed into the proper containers. For example, they expect sharps to be placed into sharps containers, and therefore they neither expect nor watch for sharps that may have been tossed in the trash.

Housekeeping personnel are best protected by emphasizing to health-care workers the importance of using the proper container for each type of waste. They must realize that failure to do so endangers all others who subsequently handle the wastes.

Maintenance Workers

Maintenance workers are at risk for occupational exposures when they maintain or repair equipment that was contaminated by spills or splashes of liquid waste. They can also be exposed when they unknowingly work on something that contains infectious or other medical waste or that was used for disposal of such wastes (for example, plumbing).

Exposures can be minimized by stressing to others the importance of using proper disposal containers and methods. All spills, including those inside equipment (such as centrifuges) should be reported and cleaned up. Maintenance workers should not work on equipment that shows any signs of spills or other contamination until it has been cleaned or disinfected by those responsible.

Operators of Treatment Equipment

For operators of infectious waste treatment equipment, there is risk from handling wastes that are not packaged in suitable containers.

Latex gloves provide protection against wet wastes. There is no glove, however, that protects against needle sticks. Staff training for the waste discarders is essential to ensure proper packaging of the wastes that these workers subsequently load into incinerators or steam sterilizers or other treatment equipment.

Trash Haulers

When they pick up the wastes from the generators, trash haulers are also at risk for exposure that results from improperly packaged infectious and other medical wastes. They can be exposed to infectious agents or chemicals from leaking waste containers. Another hazard is needle sticks from needles that were improperly contained and that protrude through cartons.

Landfill Workers

Landfill workers are at risk for exposure to wastes that are sent to the landfill for disposal. Federal regulations prohibit the disposal of hazardous and toxic chemicals and of radioactive wastes in sanitary landfills.* Although there are no regulations on the federal level prohibiting landfill disposal of untreated infectious waste, some states do have such prohibitions (although most do not).

As a result, infectious wastes are sometimes sent to landfills for disposal. At the landfill, bulldozers compact the waste as part of the burial procedures. Such compaction breaks open the containers, creating dusts and aerosols. Landfill workers are at risk for exposure to infectious agents in the untreated waste, especially in the dusts and aerosols. Equipment operators should routinely work in ventilated cabs because of the dust conditions and the accompanying risk of exposure to infectious agents.

Sharps in the waste are a special cause for concern because of the double hazards of injury and disease. Another problem with sharps is that they become stuck in the wheels and tractors of the bulldozers and other landfill vehicles; the sharps are then a hazard to those who perform maintenance work on the equipment.

PRUDENT PRACTICES FOR RISK REDUCTION

A good and comprehensive waste management system can minimize the occupational safety and health risks that are associated with the handling of infectious and other medical wastes. These risks must be considered when policies and procedures for waste management are being established. It is important to adopt prudent practices that minimize the occupational risks of waste handling.

The procedures that are listed in Table 14.1 as prudent practices were discussed in detail in other chapters of the book. They are presented here again in the context of practices that are effective in reducing occupational risks.

OTHER MANAGEMENT TECHNIQUES FOR RISK REDUCTION

In addition to adopting prudent practices for waste management procedures, there are other management techniques that can reduce the occupational risks associated with the handling of infectious and other medical wastes. These include immunizations against certain diseases, job training, and preparedness for emergency response.

---

*Regulations promulgated by the U.S. Environmental Protection Agency under the Resource Conservation and Recovery Act and the Toxic Substances Control Act (TSCA) control the disposal of hazardous and toxic chemicals. Regulations promulgated by the Nuclear Regulatory Commission control the disposal of radioactive wastes.

Table 14.1.   Prudent Practices for Risk Reduction

Procedure	Risk Reduction
Waste Identification	
Classify wastes in order to identify those that require special management and handling.	Defines waste categories in accordance with official policy.
Waste Separation and Segregation	
Separate wastes that require special handling (that is, infectious, hazardous, and radioactive wastes) from those that do not.	Separates out those wastes that require special handling. Removes them from the general trash.
Use distinctive containers for the special wastes.	Makes special wastes easily recognizable.
Discard wastes directly into designated containers (no subsequent sorting of wastes).	Minimizes waste handling and possibilities of exposure.
Segregate the different special waste streams (that is, the infectious, hazardous, and radioactive).	Reduces errors by not commingling the different types of waste. Promotes separate handling of the special waste streams.
Waste Collection and Movement Within the Facility	
Use appropriate personal protective equipment (such as gloves, masks, goggles, aprons) when handling special wastes.	Provides protection against exposure.
Use caution in lifting and moving waste containers. Use correct techniques for lifting heavy objects.	Reduces the likelihood of back and other injuries.
Use appropriate carts for collecting and moving the different kinds of waste containers.	Reduces the possibility that waste containers will fall off the cart, thereby reducing the incidence of injury and exposure.
Use carts that are easy to load and unload, to move and maneuver, to brake, and to clean.	Reduces the incidence of accidents and spills and, therefore, of injury and exposure.
Close and seal (as appropriate) all waste containers before they are moved.	Minimizes exposures.
Waste Storage	
Use storage areas that are suitable to the type of waste being stored.	Minimizes accidents, spills, and exposures.
Use storage areas that are limited to the storage of waste.	Minimizes the possibility of inadvertent use of wastes and contamination of materials.

Table 14.1, continued

Procedure	Risk Reduction
Maintain accurate records that identify each container, its contents, date of placement into storage, and (for radioactive wastes) the time required for decay of the radioisotope.	Ensures timely movement of wastes out of storage, thereby reducing possibility of exposure to deteriorating wastes or containers. Prevents premature release from storage of radioactive wastes, thereby reducing possibility of exposure to radioactivity.
Limit access to authorized personnel.	Minimizes the possibility of exposure for persons who might unknowingly handle special wastes.
Keep areas clean and vermin-free.	Prevents vector-borne transmission of infectious agents.
Properly post or otherwise identify storage areas.	Provides ready identification for emergency response teams.
Waste Treatment	
Make sure operations are performed only by adequately trained personnel.	Minimizes risk of exposure for persons mishandling wastes or equipment.
Use personal protective equipment as appropriate.	Minimizes exposures.

Immunizations

Immunizations can prevent certain diseases. Therefore, they reduce the risk of occupationally acquired disease in waste handlers. These immunizations should be made available to all personnel who handle infectious waste. In fact, it may be best to adopt a policy that requires immunizations for all persons who handle infectious and medical wastes. Immunizations that are appropriate for waste handlers include hepatitis B vaccine, tetanus vaccine, and gamma globulin.

Job Training

The best procedures are worthless if personnel are not trained in their use. The following steps are essential for implementation of a waste management system:

1. Adopt policies.
2. Establish procedures that reflect the adopted policies.
3. Develop standard operating practices (SOPs) for the established procedures.
4. Formalize the SOPs in written form.
5. Train personnel in using the SOPs.
6. Make spot checks to verify that established procedures are being followed.

See Chapter 16 for a detailed discussion of job training, teaching techniques, and the training program.

Response to Exposures

Even with the best waste management system, there can be exposures to the hazards in wastes. Of course, chances of exposure are much greater with a less-than-optimum system. Exposures can result from accidents, carelessness, and inadequate training, as well as from use of improper procedures, inappropriate waste containers, and unsuitable waste handling and treatment equipment.

It is essential that there be a program for response to exposures. Such a program should include first aid, additional medical care as necessary, medical surveillance as advisable, exposure reports, recordkeeping, and incident evaluation. In the interests of employee health and safety, the exposure reporting should be nonpunitive to encourage prompt, accurate, and full reporting of all incidents.

The appropriate medical response to exposures depends on the type of exposure: how the person was exposed and to what. The initial response must be first aid, such as cleansing of wounds and flushing of eyes or skin. Evaluation of the exposure may indicate the advisability of additional steps including inoculations, blood tests to establish baseline levels, and surveillance to determine any developing medical condition.

Needle sticks are of particular concern at present because of the possibility of infection with the HIV virus and subsequent development of AIDS. Wound cleansing should be followed with clinical evaluation and additional treatment (such as gamma globulin and antibiotics) as appropriate. The medical program should include immediate and follow-up testing for HIV antibodies, screening for HBV infection, and health surveillance.

The CDC is sponsoring a prospective study to determine the incidence of HIV infection in health-care workers exposed to the blood of infected patients.* For this study, initial blood tests are performed post-exposure, with periodic follow-ups to determine seroconversion. (You should consider enrolling exposed workers in this study. Contact the CDC AIDS Needlestick Surveillance Study Office at (404)639-3406 for more information.)

Response to Accidents

The immediate response to accidents must be first aid to the injured employee. When the accident involves spills of infectious or other medical wastes, emergency response is needed for spill containment and clean-up in order to limit exposures. (This section deals with the medical response. Emergency response is discussed in detail in Chapter 15.)

---

*In a prospective study, exposed workers are followed over a period of time in order to determine health effects of the exposure. A retrospective study attempts to correlate present disease with past exposure.

The program for response to accidents should include first aid for accidents and exposures (see above), follow-up medical care, and health surveillance as appropriate. In addition, accident reports, recordkeeping, and accident evaluation are also important elements in the accident response program.

Accident analysis is essential for evaluating the safety of the waste management system. Such evaluation can indicate the need for changes in procedures or equipment that could safeguard workers by reducing the incidence of future accidents. The reporting of accidents must be nonpunitive to encourage prompt and accurate reporting of all accidents so that the record will provide an accurate and complete database for accident analysis.

REFERENCES

1. U.S. Department of Labor and U.S. Department of Health and Human Services. "Joint Advisory Notice: Protection Against Occupational Exposure to Hepatitis B Virus (HBV) and Human Immunodeficiency Virus (HIV)." October 19, 1987. (Also in *Federal Register* 52(210):41818–41824, October 30, 1987).
2. Brock, W. E., and O. R. Bowen. Letter of October 30, 1987 addressed to "Dear Health-Care Employer" that accompanied the mailing of Reference #1. Also published in *Federal Register* 52(210):41818, October 30, 1987.
3. Department of Labor, Occupational Safety and Health Administration. "Occupational Exposure to Bloodborne Pathogens; Proposed Rule and Notice of Hearing," *Federal Register* 54(102):23042–23139, May 30, 1989.
4. Centers for Disease Control. "Recommendations for Prevention of HIV Transmission in Health-Care Settings," *Morbidity and Mortality Weekly Report* 36(2S):1S–18S, August 21, 1987.
5. *Code of Federal Regulations* Title 10, Part 20.202. "Personnel Monitoring."
6. Joint Commission on Accreditation of Healthcare Organizations. Standard #NM.2.2.16, in chapter on Nuclear Medicine Services, in *Accreditation Manual for Hospitals,* 1990 ed. (Chicago: JCAHO, 1989).
7. Joint Commission on Accreditation of Healthcare Organizations. Standard #RA.2.2.12, in chapter on Radiation Oncology Services, in *Accreditation Manual for Hospitals,* 1990 ed. (Chicago: JCAHO, 1989).
8. U.S. Environmental Protection Agency. "Standards for the Tracking and Management of Medical Waste; Interim Final Rule and Request for Comments," *Federal Register* 54(56):12326–12395, March 24, 1989.

CHAPTER 15

Preparing for Hazardous Material Emergencies

Hazardous* material emergencies can occur anywhere a hazardous material is used, including facilities that generate infectious and medical waste. Infectious agents, chemicals, and radioactive materials have all initiated emergency incidents, such as unplanned releases that threaten the environment and exposures that may harm employee health. Oxidizing, flammable, corrosive, or explosive chemicals can damage property and cause injury. This chapter addresses hazardous material emergencies that can occur at hospitals, clinics, schools, medical facilities, and other institutions that handle hazardous material of low to moderate risk. Industrial and specialized facilities that have risks beyond the scope of this discussion may need to consult further with experts, although the following discussion will provide a good background.

Planning for hazardous material emergencies involves a systematic approach that includes anticipating possible incidents, making a reasonable effort to prevent their occurrence, and preparing responses to any such incidents that may occur. This chapter addresses all of these planning elements while making specific suggestions for the content of a contingency plan and the steps to take to clean up a spill of infectious, chemical, or radioactive material.

COMMUNITY EMERGENCY PLANNING

Medical facilities have long been vital participants in community planning and response to hurricanes, floods, tornadoes, fires, and other mass casualty

---

*In this chapter, the terms "hazardous material" and "hazardous" are used broadly and generically to include chemicals, radioactive materials, and infectious agents.

incidents. Hospital personnel may also have to cope with injury to staff and patients in the case of an onsite fire or severe weather.

Plans for regional nuclear disasters are well developed and the framework for community chemical emergency planning has recently been established. The role of medical institutions in community emergency response is to offer care and advice. In some cases the local hospital is relied on to provide communication to relatives, neighbors, media personnel, and emergency response officials.

Chemical Emergency Planning

Title III of the 1986 Superfund Amendments and Reauthorization Act (SARA) requires communities to establish chemical emergency response plans. Although participation is not compulsory, the National Response Team has strongly recommended including representatives of local health departments, hospitals, emergency medical services, veterinarians, and the medical community in preparing the plan.[1]

Medical institutions should offer their facilities and expertise to assist in this effort. By participating in contingency planning, the resources and limitations of the local medical community can be better understood. For example, few medical care facilities have equipment or properly trained personnel to decontaminate victims of a hazardous material release. Through planning the need for decontamination facilities can be recognized and corrected.

Also determine how the planners foresee each participant's role in chemical emergency response, what is expected of the institution, and how the plan is coordinated with other health care facilities. It is best for each institution to designate an employee to coordinate emergency response planning, to review plans for responses that the institution has agreed to participate in, and to maintain records (e.g., plans and agreements of assistance).

SARA Requirements for Medical Institutions

As directed by SARA, the EPA promulgated regulations addressing chemical emergency response. Facilities (including generators of medical and infectious waste) that use any listed extremely hazardous substance in quantities above the specified threshold quantity are required to notify the Local Emergency Planning Committee (LEPC), the State Emergency Response Commission, and the local fire department (see Figure 15.1[2]). A facility emergency coordinator must also be designated.

In general, the aggregate total quantity of chemicals used in most hospitals and other medical institutions is below the threshold quantities (see Table 15.1). However, facilities that have large research programs should review their stocks of these chemicals and check purchasing records, if available, to determine if they need to comply. Also, facilities that store more than 1000 lb of ethylene oxide (a gas sterilant) would be subject to emergency planning notification

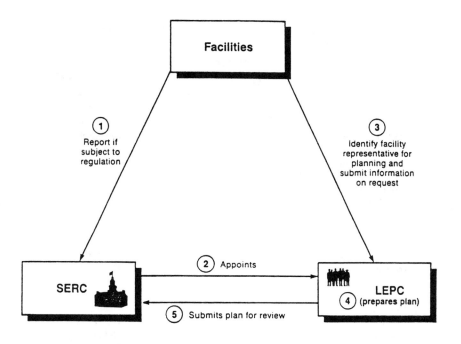

SERC = State emergency response commission

LEPC = Local emergency planning committee

Figure 15.1.   SARA Title III planning steps.

requirements (see Tables 15.1 and 15.2). (Also review 40 CFR 355.30(e) regarding mixtures; ethylene oxide is often supplied as a gas mixture.) The complete list of extremely hazardous substances is given in Appendix A of 40 CFR 355.

Medical facilities and research laboratories are exempt from additional SARA reporting requirements (see Table 15.2), as are institutions that are not subject to the Occupational Safety and Health Administration's Hazard Communication Standard (e.g., state-operated institutions not subject to separate state regulations). Note that state and local laws may be more stringent than those under SARA, and may have lower thresholds and fewer exemptions. Contact the LEPC, local fire department, and State Emergency Response Commission to determine the exact local requirements for chemical emergency planning and community right-to-know in your area. In any case, an exemption from emergency planning does not mean that a chemical hazard does not exist at the facility, or that contingency planning is unnecessary. Instead, the regulations provide some exemptions from

Table 15.1.   Extremely Hazardous Substances Potentially Present at Large Research
Institutions[a,b]

Chemical Name	Threshold Planning Quantity (lb)	Method of Use
Acrolein	500	Research
Acrylamide	1000	Research
Ammonia	500	Refrigeration
Cadmium oxide	100	Research
Chloroacetic acid	100	Research
Chromic chloride	2	Research
Dimethyl-P-phenylenediamine	2	Research
Dinitrocresol	2	Research
Ethylene oxide	1000	Sterilant
Fluoroacetyl chloride	2	Research
Hydroquinone	500	Research
Mercuric chloride	500	Research
Mercuric oxide	500	Research
Methyl disulfide	100	Research
Methyl vinyl ketone	2	Research
Nickel carbonyl	Any	Research
Nitric acid	1000	Research
Phenol	500	Research
Phosgene	2	Research
Phosphorus pentoxide	2	Research
Potassium cyanide	100	Research
Propargyl bromide	2	Research
Sodium azide [Na(N$_3$)]	100	Research
Sodium cyanide [Na(CN)]	100	Research
Sulfur dioxide	500	Research
Sulfuric acid (heating plant)	1000	Research
Titanium tetrachloride	100	Research
Toluene 2,4-diisocyanate	500	Research
Toluene 2,6-diisocyanate	100	Research
Trichlorophenylsilane	2	Research
Trimethylchlorosilane	1000	Research
Vanadium pentoxide	100	Research
Vinyl acetate monomer	1000	Research

[a]This is an abbreviated list from Appendix A of 40 CFR 355.
[b]Substances that are used in research labs and medical facilities are exempt from reporting requirements (40 CFR 370.2) unless present in amounts greater than threshold planning quantities.

a few regulatory requirements but specify that prevention and planning are the
responsibility of individuals who use hazardous materials.

Contingency Plans for Chemical Waste

As explained in Chapter 10, EPA requires contingency planning for all institutions generating more than 100 kg/month of regulated chemical waste. These

Table 15.2.   Facility Requirements Under SARA[a]

Type of Facility	Hazardous Chemicals Present[b]		
	Any Amount	>500 lb	EHS >TPQ
Research laboratories and medical facilities[c]	Exempt	Exempt	Planning[d]
Other facilities	MSDS[d]	Inventory[d]	Planning, inventory

[a]Requirements are based on the type of facility and the amounts and identity of materials on hand.
[b]Hazardous chemical is defined broadly in 29 CFR 1910.1200(c). EHS (extremely hazardous substance) and TPQ (threshold planning quantities) are both listed in Appendix A of 40 CFR 355. There are some other exemptions.
[c]Pertains to activities conducted under the supervision of a technically qualified individual. Custodial and grounds activities at medical institutions, for example, may be treated the way other facilities are.
[d]*Planning* requires notification of the Local Emergency Planning Committee, State Emergency Response Commission, and local fire department. *Material Safety Data Sheet* (*MSDS*) reporting to the above three agencies must be done annually, specifying all the hazardous chemicals at the facility. *Inventory* reporting to the same agencies must detail use and storage.

requirements are published in the *Code of Federal Regulations* [40 CFR 262.34(d)(5)] and reprinted below:

(i) At all times there must be at least one employee either on the premises or on call (*i.e.*, available to respond to an emergency by reaching the facility within a short period of time) with the responsibility for coordinating all emergency response measures specified in paragraph (d)(5)(iv) of this section. This employee is the emergency coordinator.

(ii) The generator must post the following information next to the telephone:
(A) The name and telephone number of the emergency coordinator;
(B) Location of fire extinguishers and spill control material, and, if present, the fire alarm; and
(C) The telephone number of the fire department, unless the facility has a direct alarm.

(iii) The generator must ensure that all employees are thoroughly familiar with proper waste handling and emergency procedures, relevant to their responsibilities during normal facility operations and emergencies;

(iv) The emergency coordinator or his designee must respond to any emergencies that arise. The applicable responses are as follows:
(A) In the event of a fire, call the fire department or attempt to extinguish it using a fire extinguisher;
(B) In the event of a spill, contain the flow of hazardous waste to the extent possible, and as soon as is practicable, clean up the hazardous waste and any contaminated materials or soil;
(C) In the event of a fire, explosion, or other release which could threaten human health outside the facility or when the generator has knowledge that a spill has reached surface water, the generator must immediately

notify the National Response Center (using their 24-hour toll free number 800/424-8802). The report must include the following information:

(1) The name, address, and U.S. EPA Identification Number of the generator;
(2) Date, time, and type of incident (*e.g.*, spill or fire);
(3) Quantity and type of hazardous waste involved in the accident;
(4) Extent of injuries, if any; and
(5) Estimated quantity and disposition of recovered materials, if any.

Required contingency planning is more formal and detailed for generators of more than 1000 kg/month of regulated chemical waste or more than 1 kg/month of a listed acute hazardous waste.

Emergency Planning for Infectious Agents and Radioactive Materials

Plans for emergencies involving infectious agents and low-level radioactive materials are not as strictly regulated on the federal level as are those involving chemicals. The Nuclear Regulatory Commission (and state agencies in agreement states) specifies emergency response in the license conditions of institutions that use radioactive matierals. For infectious agents, emergency planning is usually framed by the guidance and standards issued by accrediting bodies for health care facilities. In addition, extensive voluntary planning and response efforts by institutions that use these materials have adequately addressed emergency risks.

HAZARD ANALYSIS

Anticipating emergencies is an exercise in asking, "What if?" What if a bag of infectious waste rips open on the loading dock? What if a radioactive source gets lost in the laundry? What if there's a spill in the lab? Of course, the number of possible accidents is countless. The best use of resources requires analysis of possible emergencies.

Risk analysis systematically ranks events by their probability and severity. The aim is to identify those incidents that are most likely to occur and those with the potential to cause the greatest harm (see Figure 15.2[2]). By definition, emergency planning focuses on low-probability risks. Note that highly probable emergencies of medium severity and high-severity emergencies of medium probability also fall within the scope of concern for contingency planning.

With hazardous materials, severity is relative to the potential degree of harm. Infectious agents that cause fatal and debilitating diseases typically call for serious control measures, even if the probability of transmission is small. Chemical carcinogens, teratogens, and mutagens have been targeted for additional protective

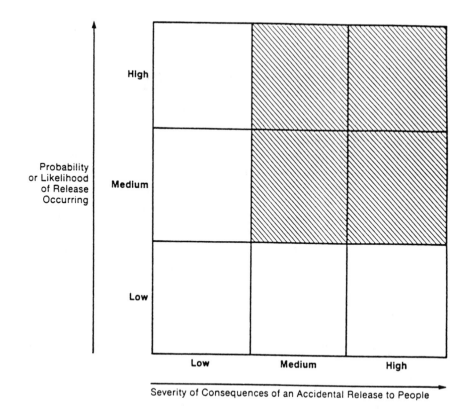

Figure 15.2.   Risk analysis matrix.

measures among the hundreds of thousands of chemicals that exhibit other forms of toxicity.

For administrators and risk managers, severity is measured in terms of cost. Hazardous material accidents can result in employee time spent cleaning up a spill, disposal fees, sick leave and workers' compensation claims from injured employees, fines for releases and improper management, and possibly litigation costs. A hazardous material incident can also have a political cost, dimming the reputation of an institution within the community or straining relations with employees. The institution's chief executive officer has to balance these potential costs when allocating monies for planning, prevention, and preparedness.

The severity of an emergency can be manifested by damage to human health, the environment, or property. The potential of flammable materials to cause property damage is widely recognized and preventive measures are in place. The potential environmental impact of a hazardous material release, however, is difficult to assess. Even a small quantity of a spilled organic solvent can lead to groundwater contamination. Uncontained transport of infected livestock and carcasses (not an uncommon practice) can lead to a regional epidemic.

Occupational risks of hazardous materials are widely acknowledged, but their perception and control measures have progressed steadily as we understand more about toxicity and disease. Much remains unknown about the toxicity of most chemicals. The study of toxicology is complicated by the insidious nature of chronic low-level exposure.

In addition to degree of hazard or toxicity, the quantity or concentration of the material is a factor in assessment. Chemical and radiation injuries are dose-dependent. In some cases, there may be low-level threshold concentrations, below which injury does not occur. For infectious agents, an exposure below the infectious dose will not result in disease.

Diseases attributed to toxic chemicals and infectious agents are the result of direct contact of the chemical or agent with the target organism (e.g., man, fish, etc.). Direct contact (e.g., ingestion, skin absorbtion) is also a hazard of many radioactive materials used in health care, although radiation can cause harm even when the radioactive material itself is contained. Still, containment is the primary goal of prevention of, response to, and cleanup of hazardous material spills. Because emergencies are typically sudden events, the effect of a short-term exposure (e.g., acute toxicity) is of great concern.

Analysis of Hazardous Materials

Begin planning by identifying the hazardous materials present at the facility, how they are stored, used, and disposed of, and their amounts and concentrations. Use a hazardous material survey for systematic risk assessment.

Consider possible routes of release. How likely is a spill or leak? Where is the fuel oil tank overflow outlet in relation to the storm sewer drain? Would some release routes have severe consequences? (For instance, many pesticides kept for pest control and grounds maintenance are toxic, but their combustion products are much more toxic in the event of a fire.) If water is used to extinguish a fire involving hazardous material, where would the contaminated water go? Is the local sanitary sewerage treatment works capable of treating such runoff?

Analysis of Practices

From a historical perspective, employee accident reports and spill notifications can provide a wealth of information for risk analysis. The probability of an incident can be estimated from the number of reports filed. Severity can be measured by a variety of means, such as work hours lost and spent attending to the incident. For hospitals, back injuries are high on the list of costly workers' compensation claims. Needle sticks may not be unavoidable; the threat of hepatitis B and AIDS calls for extensive prevention measures. The usefulness of accident reporting underlines its importance; make sure that employees know to report all hazardous material exposures and spills, as well as other types of releases and accidental exposures to workers. Revise the accident report form to ask for hours spent for response and other costs so that the information will be more useful in quantitative risk analysis.

Investigate the experiences of similar institutions. For example, many hospitals have a history of occasional mercury spills in clinical areas and laboratories where it is used to measure gas pressures. When spilled, mercury finds its way into cracks and beneath floor tiles. Over the years small amounts can accumulate and, at higher temperatures, produce hazardous vapor levels in the room. Each institution that uses mercury, even if it has had few spills, might as well prepare for such an experience.

Identify and inspect sites where hazardous materials and wastes are stored, used, collected, treated and disposed of. "What if?" planning is best done when the viewpoints of management, safety experts, and staff are considered. Be especially observant for risky practices that have become accepted through years of use. Assessment of offsite waste handling facilities is a must. Chapter 17 discusses ways to reduce risks through selection of waste contractors and disposal facilities.

Analysis of Processes and Procedures

Emergencies can result from equipment breakdowns and the failure of employees to follow procedures when handling hazardous materials and wastes. Look objectively at the management of hazardous materials to identify conditions that would result in a release or other accident. Review maintenance logs for records of equipment failure and repair. Interview maintenance staff and contractors, custodians, and refuse handlers for their recollection of releases and other problems that may have gone unreported. As discussed in Chapters 6 and 7, steam sterilizers and incinerators can release infectious agents when they are not maintained or operated properly.

In addition to acute emergencies (e.g., spills), plan for potential failures in the waste handling system that could cause a release, such as a treatment equipment upset that results in the failure to sterilize infectious waste. Nonsudden emergencies include the discovery of a slow leak of a past spill, or inadvertently disposed-of untreated waste. General breakdowns in the waste management system should also be planned for; this type of contingency planning is dealt with in Chapter 16.

PREPAREDNESS AND PREVENTION

When the areas of greatest risk are identified, incident prevention deserves to be considered before emergency planning. The process of analyzing hazards and potential emergencies usually leads to suggestions of ways to prevent and prepare for them. Prevention and preparedness can be accomplished by modification of the facility, changes in procedures, the addition of communication systems, and health surveillance of employees who may come in contact with hazardous materials. Most of these items require additional expenditures; risk analysis can help justify them and set priorities.

Engineering Controls: Safety engineering is frequently the best way of preventing or minimizing hazardous materials emergencies. Surround areas where liquids are stored by a curb to dike spills that may occur. Provide ventilation or respiratory protection for areas where vapors or aerosols of hazardous materials may be present. Stored flammable materials must have portable fire extinguishers nearby and, ideally, fire suppression systems installed to cover the area. Store infectious waste in a cool area; refrigeration and freezing are appropriate for prolonged storage. After installation, all engineering controls require periodic maintenance and testing to ensure they continue to operate properly.

Interviews with waste handlers and site visits frequently turn up complaints of inadequate equipment, maintenance, or facilities. This is the time to prepare a wish list of those maintenance items that need to be funded and equipment that should be purchased; risk analysis can justify such expenditures. Compare the engineering controls used at similar institutions either by touring other facilities or by phoning peers for their advice.

Procedural Controls: For manual operations, adopt procedures that will help to prevent accidents. Supervisors are responsible for making sure that procedures are followed and safety rules are strictly enforced including rules for use of personal protective equipment (e.g., gloves, aprons, masks, etc.; see Chapter 14 for occupational safety) and maintenance of adequate aisle space so as to not interfere with facility evacuation.

Communication System: Several routes of communication are available, both internal and external. Internal communications (e.g., alarms, intercom, phone, voice) are necessary to evacuate the area or facility when there is a threat of exposure to toxic materials. Have external systems (e.g., phones, CB radios) available to notify outside responding agencies.

Medical Surveillance and Preventive Medicine: Medical attention for employees is appropriate not only to detect changes in employee health resulting from exposures to hazardous materials, but also to prevent disease. Administration of the hepatitis vaccine and other preventive immunizations is appropriate for many workers who may be exposed to infectious agents and waste. See Chapter 14 for further discussion of medical monitoring.

CONTINGENCY PLANNING

Focus the emergency plan on several specific, identifiable incidents or groups of similar incidents. One likely scenario is infectious waste spills. It is difficult to plan for generic, nonspecific emergencies. The plan must be flexible, therefore, because not all incidents and their conditions can be anticipated. During an emergency, variations and intervening factors are inevitable; every situation is unique.

The Planning Process

Emergency planning can be a detailed, complex process (see Figure 15.3[1]). To simplify the task, remember to address only those incidents for which the facility is most at risk. Remember too that the plan will serve as a reference for training, so it may contain some background information. Don't let the plan become excessively elaborate; overplanning only results in a plan that is hard to conceptualize and that probably would remain unread.

Enlist others to help with planning. EPA suggests seeking safety advice from within the institution and, for community hazardous material planning, a diverse group of consultants (see Table 15.3).

The objective of planning for emergencies is to identify actions that need to take place in an emergency and prepare for them ahead of time, leaving as many resources as possible available for emergency response. Many steps can be completed, or in place, in anticipation of an accident:

- identifying likely incidents
- outlining emergency scenarios
- establishing command hierarchy
- organizing lines of communication
- determining response actions
- delegating responsibilities
- designating an evacuation signal and a rendezvous point

The plan must also coordinate well with the local emergency response plan. The local hazardous material plan may focus primarily on chemical emergencies, but in many cases it will be combined with the current local plan covering all types of hazards that may threaten the community (e.g., severe weather, large fires, nuclear disasters, transport accidents). The infrastructure of agencies relied on for local disaster response is the same for a variety of emergencies.

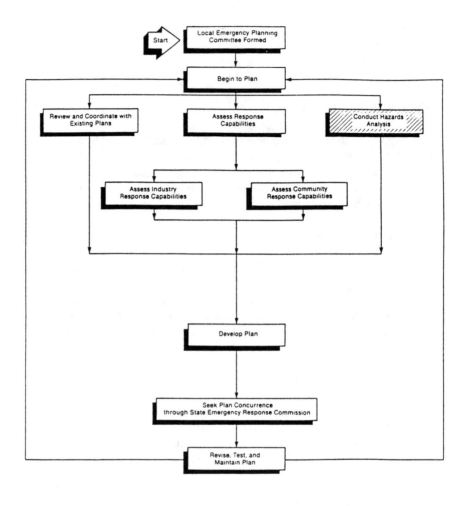

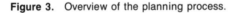

*From NRT-1

Figure 3. Overview of the planning process.

Make sure that throughout the planning process there is review by the institution's administration. The planning group should brief key officers when milestones are reached (see Figure 15.3) so that the decisions and details are understood as well as their importance. When completed, the plan must have full administrative approval so that funding for preparedness and prevention is consistent with its scope. Administration should know the extent of available response measures and the additional costs associated with them. For example, most community hazardous material (HazMat) squads charge disposal fees and for supplies used to clean up a spill. If a commercial emergency response contractor is called, costs can soar, so this should be decided beforehand.

Table 15.3.   **Potential Members of an Emergency Planning Team**

Experience shows that the following individuals, groups, and agencies
should participate in order for a successful plan to be developed:

Mayor/city manager (or representative)[a]

County executive (or representative)/board of supervisors[a]

State elected officials (or representative)[a]

Fire department (paid and volunteer)[a]

Police department[a]

Emergency management or civil defense agency[a]

Environmental agency (e.g., air and/or water pollution control agency)[a]

Health department[a]

Hospitals, emergency medical service, veterinarians, medical community[a]

Transportation agency (e.g., DOT, port authority, transit authority, bus company, truck or rail companies)[a]

Industry (e.g., chemical and transportation)[a]

Coast Guard/EPA representative (e.g., agency response program personnel)

Technical experts (e.g., chemist, engineer)

Community group representative[a]

Public information representative (e.g., local radio, TV, press)[a]

Other groups/agencies that can be included in the planning process,
depending on the community's individual priorities:

Agriculture agency

Indian tribes within or adjacent to the affected jurisdiction

Public works (e.g., waste disposal, water, sanitation, and roads)

Planning department

Other agencies (e.g., welfare, parks, and utilities)

Municipal/county legal counsel

Workers in local facilities

Labor union representatives (e.g., chemical and transportation, industrial health units)

Local business community

Representatives from volunteer organizations (e.g., Red Cross)

Public interest and citizens' groups, environmental organizations, and representatives of affected neighborhoods

Schools or school districts

Key representatives from bordering cities and counties

State representatives (Governor, legislator's office, state agencies)

Federal agency representatives (e.g., FEMA, COT/RSPA, ATSDR, OSHA)

[a]Required by Title III of SARA.

Using Risk Analysis Data

Although quantitative data are rarely available, the risk analysis exercises suggested above should provide some qualitative information for ranking hazardous material risks. A group of experts (e.g., safety director, risk manager, biosafety officer, infection control nurse, plant engineer) should review this information to objectively rank those incidents of greatest risk due to their severity and/or probability.

Decisionmaking

Perhaps the most important preparation is deciding on the administrative policies to be followed in an emergency. Decisions that need to be made include:

- Under what conditions should there be a full or partial evacuation?
- When should additional resources be requested of outside agencies?
- What should be told to local media and who should be the facility's spokesperson?

At a minimum, administration should discuss the framework of these policies ahead of time with the emergency coordinator and institutional safety professionals.

Emergency Coordinator

Who is in charge in the event of a hazardous materials emergency? Designate a facility emergency coordinator, and alternates who can take charge if he or she is not available. The coordinator is responsible for assessing the incident's hazard and choosing an appropriate response or interim control measure. The emergency coordinator must have the authority to commit the necessary institutional resources to implement the contingency plan. The local emergency response plan describes when that authority is to be surrendered during community-threatening disasters, and to whom it is surrendered.

The emergency coordinator may need access to outside experts when an incident is severe or difficult to get under control. Hazardous material emergencies can become complex, involving many kinds of occupational and environmental hazards. Additional emergency information regarding chemicals is available through various databases and CHEMTREC (1-800-424-9300). For infectious agents, this is available through the Office of Biosafety, Centers for Disease Control (1-404-633-3311, Ext. 3883). Along with local and institutional experts available to the emergency coordinator in an emergency, these resources can meet the need for ongoing hazard assessment in extended incidents.

Procedures

Establishing written emergency procedures and designating responsibilities is perhaps the most difficult aspect of contingency planning because it requires the consensus of so many groups. Cover each of the following tasks:

- notification of staff, administration, and the emergency coordinator: always notify the local responders (i.e., fire or police department), if only to keep them posted as to the progress of cleanup.
- prevention of further exposure and release by containing the hazardous material

- securing the area
- cleaning the area and removal or containment of the hazardous materials to safe exposure levels

Include sufficient details and specific responsibilities by job title to ease their implementation and allow their use in staff training. Also draft procedures to ensure readiness; specify who is responsible for checking emergency response supplies and equipment.

Communications

Outline the purpose of each communication system and who would have access in an emergency. Provide for a command post that could serve as a communications center and where decisions can be made without undue distractions. Choose a command post location that is distant from the scene of an incident, but with good access to communication, such as the institution's security office.

Institutional HazMat Squad

Facilities may choose to select a group of employees for special hazardous material training and give them exclusive response responsibilities. In the event of a spill, these employees are paged to report to the command post for instructions from the emergency coordinator. Such a team allows for additional, customized training, including detailed knowledge of material use and hazards and proper use of respirators, as well as medical surveillance of team members.

Arrangements with Local Authorities

When the magnitude and severity of a hazardous material emergency exceeds the capabilities of institutional personnel, outside agencies must be called on to assist. Although the fire department is relied on to respond to even small fires, until recently few communities have had similar response units for liquid spills, gaseous releases, or accidental exposures. Many urban areas have now established hazardous material response teams (HazMat squads) that are specially trained and equipped to respond to spills of and fires involving hazardous materials. A unit of the local fire department has assumed these duties in many cases. Some government agencies have a defined role in emergency response and are required to assume control in events that threaten public health. Other local government agencies have a role in emergency response, as shown in Table 15.4.

Identification of Resources

Equipment, supplies, and assistance are necessary for emergency response. Resources necessary for responding to those incidents that you identified as

Table 15.4. Local Agencies with Emergency Response Roles[a]

Agency	Usual Response Role
Emergency management	Coordination, communication
Emergency medical technicians	Injury and trauma care, rescue
Environmental regulators	Cleanup, decontamination
Police	Security, evacuation
Public health officials	Assessment, advice
Red Cross and similar organizations	Shelter, care

[a]Severe emergencies may involve regional, state, and federal response.

probable should be maintained onsite. All facilities need a phone or other communication device where hazardous materials are routinely handled (e.g., laboratory, waste storage area). Laboratories should have spill control equipment. For low-probability incidents, community-wide planning provides a mechanism to pool and share resources.

Legal Requirements

In many cases, it is a legal requirement to report spills and releases of hazardous materials. EPA requires that spills of hazardous materials in quantities greater than a specified reportable quantity be reported to the National Response Center (1-800-424-8802). The hazardous materials covered and their reportable quantities are given in Table 302.4 of 40 CFR 302. Requirements for emergencies involving hazardous chemical waste read: "In the event of a fire, explosion or other release which could threaten human health outside the facility or when the generator has knowledge that a spill has reached surface water, the generator must immediately notify the National Response Center" (40 CFR 262.34(d)(5)(iv)(C)).

Many states, counties, and local jurisdictions have more stringent rules: be sure to check. Written reports may be required. In some areas, all spills must be reported, regardless of the material or amount. Currently, there are no federal laws that apply to a spill of infectious waste or an etiological agent.

THE CONTINGENCY PLAN

The product of contingency planning is a written contingency plan. Table 15.5 is a sample outline of a plan suggested by the federal National Response Team. The Local Emergency Planning Committee may have a suggested format so responders will be familiar with the organization of plans from different locations.

Table 15.5.   Sample Outline of a Hazardous Materials Emergency Plan

A. Introduction
    1. Incident information summary
    2. Promulgation document
    3. Legal authority and responsibility for responding
    4. Table of contents
    5. Abbreviations and definitions
    6. Assumptions/planning factors
    7. Concept of operations
        a. Governing principles
        b. Organizational roles and responsibilities
        c. Relationship to other plans
    8. Instructions on plan use
        a. Purpose
        b. Plan distribution
    9. Record of amendments
B. Emergency assistance telephone roster
C. Response functions
    1. Initial notification of response agencies
    2. Direction and control
    3. Communications (among responders)
    4. Warning systems and emergency public notification
    5. Public information/community relations
    6. Resource management
    7. Health and medical services
    8. Response personnel safety
    9. Personal protection of citizens
        a. Indoor protection
        b. Evacuation procedures
        c. Other public protection strategies
    10. Fire and rescue
    11. Law enforcement
    12. Ongoing incident assessment
    13. Human services
    14. Public works
    15. Others
D. Containment and cleanup
    1. Techniques for spill containment and cleanup
    2. Resources for cleanup and disposal
E. Documentation and investigative follow-up
F. Procedures for testing and updating plan
    1. Testing the plan
    2. Updating the plan
G. Hazards analysis (summary)
H. References
    1. Laboratory, consultant, and other technical support resources
    2. Technical library

Many plans become very detailed and complex. Such bulky documents serve well for training purposes, but are often impractical to use in an actual emergency. For on-scene responders, some communities have prepared a brochure containing a summary checklist that outlines the duties of each group.

Minimum Hazardous Waste Requirements

As discussed in Chapter 10, hazardous waste generators of 100 to 1000 kg/month have to comply with special requirements for contingency planning. These requirements represent an excellent example for low-risk users of hazardous material. The following is a summary of the requirements:

1. *Emergency Coordinator*: Assign an employee the responsibility of coordinating all emergency response measures. Make arrangements for the absence of the emergency coordinator by contacting the employee when he/she is not at work or by designating an alternate.
2. *Post Written Procedures*: Post phone numbers of the emergency coordinator and the fire department next to the telephone.
3. *Hazardous Material Training*: Train employees in safe handling of hazardous materials (to prevent accidents) and implementing the contingency plan.
4. *Emergency Procedures*: Respond to emergencies appropriately through the use of fire extinguishers and materials to contain spills. Spills must be cleaned up. If the spill threatens human health or the environment, notify the National Response Center (1-800-424-8802—a 24-hour number).

The complete contingency planning requirements for small quantity waste generators (40 CFR 262.34(d)(5)) are reprinted in the "Chemical Emergency Planning" section of this chapter.

RESPONDING TO SPILLS

For hospitals and other institutions that generate infectious and other medical waste, spills are probably the most common hazardous material emergency. Infectious, chemical, or radioactive materials or wastes may be spilled, and the same approach is used to clean up a spill regardless of whether it involves a material or a waste. Waste from spills—the spilled material, the cleanup material, and disposable items contaminated during cleanup—should be managed and disposed of appropriately as infectious, hazardous, or radioactive waste.

Take spills seriously. If they are not contained or cleaned up properly, hazardous materials can easily expose employees. Direct contact with liquids is an obvious means of exposure, but exposure can also result from inhalation or skin contact with an aerosol, vapor, or other airborne material.

During cleanup of all hazardous material spills, allow only those wearing the proper personal protective equipment to enter the area until it is decontaminated. Gloves, as well as other equipment that gets contaminated, should be changed frequently during cleanup. Anyone who has been exposed to the spilled or airborne material, either as a result of being present when the spill occurred or during cleanup, should wash thoroughly after the cleanup, especially all exposed skin.

The same general approach and procedures for spill cleanup are valid for all spills. An outline of recommended procedures is presented in Table 15.6. Spills of infectious, chemical, and radioactive materials/wastes are discussed more specifically in the following sections. See Table 15.7 for recommended supplies for cleanups of the different types of spills.

Spill kits should be readily available for use. There should be written cleanup procedures for the different types of spills, and personnel should be trained in these procedures. For certain more dangerous spills, cleanup should be delegated to designated, specially trained personnel who may serve a certain area or the entire facility.

Spills of Infectious Material or Waste

There is usually a substantial difference between a spill of infectious waste in a hospital and that of infectious material in a research laboratory. The difference is attributable primarily to the infectious agents present and their concentration and virulence.

Spills of infectious waste usually require only cleanup and decontamination of the spill area. Evacuation of the area is rarely necessary, although it is prudent to prevent exposure of additional personnel. Spills of blood and body fluids should be managed in accordance with the universal precautions that are recommended for use by health care personnel.[3]

For spills of infectious materials, it is important to determine the type of infectious agent involved in the spill, because some may require immediate evacuation of the affected area whereas others are less hazardous and evacuation would not be necessary. The Office of Research Safety of the National Cancer Institute[4] stresses that "attention to the immediate personal danger overrides maintenance of contaminant." In other words, with certain infectious agents it is essential to evacuate and secure the area before attempting to stop a spill from spreading, that is, to act first to prevent personal exposure.

Spill kits are available commercially, or they can be assembled (see Table 15.7). Appropriate containers for the spilled material and the contaminated cleanup materials are red plastic bags for most solid waste and special sharps containers for spilled sharps as well as broken glass.

Bleach is the disinfectant of choice for spills of infectious waste and material because it is effective, inexpensive, and readily available. Bleach is an aqueous

Table 15.6.   General Procedures for Spill Cleanup

1. Evacuate the area immediately (if necessary to prevent exposure of additional persons to a particularly toxic or virulent agent).
2. Provide immediate medical treatment to those exposed (if warranted by the nature of the exposure).
3. Determine the nature and the extent of the spill—what has been spilled (i.e., the chemical or biological agent), its concentration, quantity, and location.
4. Secure and post the spill area to prevent additional exposures and spread of the spill.
5. Don appropriate personal protective equipment:
    a. Always: glasses, gloves, lab coat or apron, shoe coverings.
    b. As appropriate (depending on the nature of the spill): face shield or goggles, respirator, boots.
6. Contain the spill (e.g., by diking or ringing with absorbent material).
7. Decontaminate the spilled material immediately, if so warranted (i.e., if prudent to decontaminate the spilled material before it is picked up).
8. Pick up the spilled material:
    a. Solids:
        1. Pick up by mechanical means (e.g., pan and brush, forceps).
        2. Discard as infectious, hazardous, or radioactive waste.
    b. Liquids:
        1. Absorb the spill.
        2. Discard as infectious, hazardous, or radioactive waste.
    c. Broken glass and other sharps:
        1. Pick up by mechanical means (e.g., forceps, pan and brush), *never* by hand.
        2. Discard as sharps.
9. Decontaminate the area and absorb.
10. Rinse the area (if necessary) and absorb.
11. Clean the area (if necessary) and absorb the cleaning compound.
12. Rinse the area (if necessary) and absorb.
13. Dispose of cleanup materials:
    a. Discard disposable items as infectious, hazardous, or radioactive waste.
    b. Decontaminate reusable items (such as dust pans, brushes, forceps).
14. Remove personal protective equipment.
    a. Discard disposable items as infectious, hazardous, or radioactive waste.
    b. Decontaminate reusable items (such as heavy rubber gloves, boots, aprons, gowns) before cleaning or laundering.
15. Wash thoroughly all exposed skin.
16. Obtain medical treatment and follow up as appropriate for type of exposure.

solution of sodium hypochlorite; it contains chlorine and is a strong alkaline oxidizer. Microorganisms are destroyed by bleach. Its effectiveness as a disinfectant is chiefly due to its oxidizing properties; free chlorine in the solution also contributes to the process. When bleach comes into contact with oxidizable materials, it reacts with them and is thereby deactivated. (Oxidizable materials include microorganisms, proteins, blood cells, cloth, paper, and shoes). Therefore,

Table 15.7. Summary of Spill Response Steps and Supplies[a]

Step	Supplies Needed
1. Controlling the spread (diking)	• Absorbent toweling (e.g., paper towels*), spill booms, or a circle of another absorbent
2. Treatment (optional)	• For infectious agents: use full-strength bleach if treatment must precede steps 3 and 4. • For acids: sodium or calcium carbonate, or sodium bicarbonate; for bases: citric acid powder • For other chemicals: see references 7 and 8.
3. Absorption	• Paper towels,* spill pillows, oil dry, cat litter, calcium bentonite, diatomaceous earth, or vermiculite
4. Recovery, containment for disposal, and cleanup	• Plastic bag, jar, bottle, jug, or pail* • Forceps, broom, dust pan,* mop and bucket* • For infectious agents: disinfecting detergent* • For mercury: vacuum pump or mercury sponge
5. Decontamination	• For infectious agents: 1:100 bleach solution* • For mercury: HgX (commercial product) • For other toxic chemicals: use a suitable solvent
6. Disposal	• For infectious agents: see Chapters 4–8. • For chemicals: see Chapter 11. • For radioactive materials: see Chapter 12.

[a] Personal protective equipment for all steps:
   Eye protection: safety glasses,* goggles, face shield
   Respiratory protection: ventilate (open windows,* use fans), evacuate if in doubt, use respirators if you are trained
   Skin protection: lab coat,* apron, gloves,* forceps,* boots or shoe covers
* = Minimum spill control supplies that should be on hand (spill kits are available commercially).

it is necessary to use an excess amount of bleach to ensure that all the infectious agents in the spill are destroyed. Bleach also is deactivated on storage, and so bleach solutions that are set aside for spill response must be kept fresh. Liquids present in the spill will dilute the bleach, and a more concentrated bleach solution should be used in these circumstances.

For most spills, a fresh 1:100 dilution of bleach will disinfect most surfaces once the spill area has been cleaned of oxidizable matter and excess liquid.[5] Spray the area with the 1:100 bleach solution and allow it to air dry. When highly infectious agents are present, it may be prudent to treat the spill before attempting removal and cleanup of the spilled material and liquids. Because the bleach may be completely deactivated by the spilled material, use a 1:10 dilution or full-strength bleach and a 15-minute contact time to ensure destruction of the infectious agents. For spills that may be contaminated with the agent causing Creutzfeldt-Jakob disease, undiluted bleach is recommended for disinfection.[6]

Use bleach carefully. It can injure skin on contact. Also, the fumes from undiluted bleach are hazardous, and respiratory protection may be needed.

For some spills, another disinfectant may be more appropriate than bleach. Contact the infection control staff or institutional biosafety officer for advice. Also see Table 8.2.

All waste materials from the cleanup of spills of infectious materials or wastes should be managed and disposed of as infectious waste. (See Chapters 4 through 8.) This includes the spilled material, any absorbent material, and all contaminated items that are being discarded.

Chemical Spills

The general cleanup procedures in Table 15.6 are appropriate for managing chemical spills. Table 15.7 has some more specific information that is relevant to chemical spills.

Chemical spill response must be adapted to the particular characteristics of the spilled chemical. Detailed procedures have been developed for the cleanup and disposal of many hazardous chemicals.[7,8] There are also procedures for many of the antineoplastic drugs used in health care facilities.[9]

Many chemicals are volatile, and it is essential to consider volatility when responding to a chemical spill. When volatile hazardous chemicals are spilled, it may be prudent to evacuate the affected area, to secure and post the area to prevent entry, to deny entry to anyone without proper respiratory protection, and to delay reentry of others until air concentrations have returned to safe levels.

Be sure to use personal protective equipment that is proper for the type of chemical that was spilled. Note that there are different types of respirators, and the appropriate respirator must be used for the particular chemical. Also, gloves, aprons, boots, and shoe coverings are manufactured from different materials, some of which provide protection against one type of chemical but not another. (See Chapter 14 for additional information about personal protective equipment.)

All waste materials from the cleanup of spills of chemicals that are hazardous per RCRA regulations must be managed and disposed of as hazardous waste. (See Chapter 11.) This includes the spilled material (if it is discarded), the absorbent material, and all contaminated items that are being discarded.

Procedures for Spills of Radioactive Material and Waste

The spill cleanup procedures in Table 15.6 and the cleanup supplies listed in Table 15.7 are also relevant to spills of radioactive material and waste.

The spill of a radioisotope should always trigger evacuation of the affected area and securing and posting of the area to prevent entry. Reentry should be permitted only for authorized personnel who are properly equipped with personal protective equipment.

All waste materials from the cleanup of spills of materials or wastes that are radioactive must be managed and disposed of as radioactive waste. (See Chapter 12.) This includes the spilled material (if it is discarded), any absorbent material, and all contaminated items that are being discarded.

TRAINING

Meetings and all the paper they generate do not ensure a practical plan that is usable in an emergency. Even when a plan exists, many hazardous material emergencies are poorly handled because the plan is not followed. Like all procedures, employees must be trained to implement the emergency response plan. Training may consist of supervisory instruction, posters where hazardous materials are present, an annual review of the plan with staff and participating responders, and/or exercises that simulate an incident and response. Exercises are not only a means to train responders; they can identify weak parts of the plan that need improvement.

At the most basic level, conduct a table-top exercise (i.e., a meeting with personnel who may be involved in a response to discuss emergency roles and actions) with a scenario and sequence of events being discussed by the various responding departments and outside agencies. A full-blown exercise may involve the dramatization of an accident and the employment of the actual equipment, communication systems, and spill response supplies designated by the plan.

See Chapter 16 for more information on employee training.

AFTER-ACTION REPORTS

After training and after an incident that required implementation of the plan, prepare a report to evaluate the plan's effectiveness and recommend improvements and issues to address in future training exercises. Ask others to comment as well (e.g., staff, government agencies).

All this is a lot of work. Meetings and paperwork alone could consume a career. Be aware that the remedy for a hazardous materials accident can consume a great deal of time, but don't spend a disproportionate effort on moderate risks.

It must be acknowledged here that most plans are useless while an emergency is taking place. They are written to document findings and decisions, and to be used for training so that staff will react proficiently in an emergency. The process—the discussion of viewpoints and agreement on a plan—is by itself a most valuable training exercise.

REFERENCES

1. *Hazardous Materials Emergency Planning Guide*, National Response Team, NRT-1 (March 1987).
2. *Technical Guidance for Hazard Analysis*, U.S. Environmental Protection Agency, Federal Emergency Management Agency, U.S. Department of Transportation (December 1987).
3. Centers for Disease Control. "Perspectives in Disease Prevention and Health Promotion. Update: Universal Precautions for Prevention of Transmission of Human Immunodeficiency Virus, Hepatitis B Virus, and Other Bloodborne Pathogens in Health-Care Settings." *Morbidity and Mortality Weekly Report,* 37(24):377–388 (1988).
4. Office of Research Safety, National Cancer Institute. "Laboratory Safety Monograph," U.S. Department of Health, Education, and Welfare (1979).
5. Centers for Disease Control Summary. "Recommendations for Preventing Transmission of Infection with Human T-Lymphotropic Virus Type III/Lympha Denopathy-Associated Virus in the Workplace." *Morbidity and Mortality Weekly Report,* 34:681–695 (1985).
6. Jarvis, W. R. "Precautions for Creutzfeldt-Jakob Disease," *Inf. Control* 3:238–239 (1982).
7. Armour, M. A., R. A. Bacovsky, L. M. Browne, P. A. McKenzie, and D. M. Renecker. "Potentially Carcinogenic Chemicals Information and Disposal Guide" (Edmonton, Alberta: University of Alberta, 1986).
8. Armour, M. A., L. M. Browne, and G. L. Weir. "Hazardous Chemicals Information and Disposal Guide," 2nd ed. (Edmonton, Alberta: University of Alberta, 1984).
9. Armour, M. A. "How to Chemically Dispose of Small Quantities and Spills of Antineoplastic Drugs, Hazardous Pharmaceuticals, Spills and Surplus Chemicals." Lecture presented during course on Managing Your Infectious and Hazardous Wastes, College of Engineering Professional Development, University of Wisconsin-Madison. Madison, Wisconsin, June 6–8, 1990.

CHAPTER 16

Training Staff and Waste Handlers

The training of staff and waste handlers serves two important functions. One is the training of employees in how to implement the waste management system. The other is risk communication: not only explaining the rationale for the established procedures, but also providing information on occupational risks to satisfy the employee's right to know.

Just establishing a waste management system is not sufficient. The system will not and cannot be implemented if personnel do not understand the system, and if they do not know their role, what they are supposed to do, and the importance of following standard operating practices. The best approach is to have a formal training program for employees. Such training is essential to ensure full and satisfactory implementation of the waste management system.

Institutional policies regarding employee training vary greatly. At some institutions, training is regarded as important, and ample time for training is readily available. At others, it is perceived as a nuisance, and approval for training time is difficult to obtain. It is important for everyone to realize that training programs are essential and that training is usually beneficial for the employer as well as the employees.

REGULATORY REQUIREMENTS

Hazard communication regulations (also known as "right-to-know" regulations) were issued by the Occupational Safety and Health Administration.[1,2] These regulations require that employees who use chemicals in the workplace be given information about the chemical hazards to which they might be exposed.

The federal regulations originally applied only to the manufacturing industries, but they were later extended to include nonmanufacturing industries as well.[3-5]

Many states have adopted their own OSHA regulations. As with all regulations, those at the state level must be at least as stringent as the corresponding federal regulations. Some states have expanded coverage of the right-to-know regulations by applying them to additional industries, including the health care industry. In other states, the health care industry is specifically excluded from right-to-know regulations.

From your perspective as an employer who has a personal interest in risk management, this distinction about applicability of OSHA regulations makes little difference. It is in your interest to have a knowledgeable staff—that is, personnel who are aware of the hazards in the wastes that they handle and who know what they should do to minimize and avoid occupational exposures and risks.

JCAHO REQUIREMENTS

The Joint Commission on Accreditation of Healthcare Organizations establishes standards for hospitals. Hospitals are then granted (or denied) accreditation on the basis of inspections for compliance with the established standards. The JCAHO standards for hospitals are published yearly as *Accreditation Manual for Hospitals.*[6]

Standards in the chapter on Plant, Technology, and Safety Management were revised for the 1989 edition of the Accreditation Manual. New standard #PL.1.10 requires "a hazardous materials and wastes program," while #PL.1.10.2 states that this program must include "training for . . . personnel who manage and/or regularly come into contact with hazardous . . . wastes." Standard #PL.1.10 also applies to the management of radioactive wastes.*

Although JCAHO Standard #PL.1.10 does not specifically mention infectious wastes or toxic chemical wastes, these types of wastes were cited in the previous version of the standard that required a management system for hazardous materials and wastes.[7] Therefore, the JCAHO standards for hospital accreditation require a program for the management of "hazardous" wastes including the infectious, radioactive, and toxic chemical wastes—in other words, infectious and medical wastes. The JCAHO standard requires that this waste management program include training for employees.

THE TRAINING PROGRAM

Various materials are available on preparing and presenting training programs. Some material relates specifically to right-to-know training. One example of the

---

*See reference 6. Standard #NM.2.2.11.2 for Nuclear Medicine Services and Standard #RA.2.2.10.1 for Radiation Oncology Services both require compliance with Standard #PL.1.10 and all its parts.

latter is the guidelines issued by the New Jersey Department of Health; this guidance pertains directly to education and training under the right-to-know regulations.[8] Other material relates specifically to the training of health care workers.[9]

The overall training program consists of many aspects in addition to the training materials per se. Items that you should consider when you are developing a training program include:

- course content
- hands-on training
- testing (as appropriate)
- follow-up to training
- the training schedule
- employees to be trained
- instructors and instruction
- recordkeeping

Each of these topics is discussed below.

Course Content

In developing a training program, you are naturally concerned about course content. The decision about what to include in a particular course is not simple, although sometimes there are regulations (such as those pertaining to occupational safety and health) that mandate what must be covered in a training course.

The best approach to defining course content is to decide first of all on the purpose of the course. What are the objectives of the course? Who should attend? What do you want the participants to learn? What should they know after they have completed the course?

Once you have decided on the objectives of a particular training course, you should then tailor the subject matter accordingly. The course could be based on a broad subject, on a particular topic, on the background of the participants, or on techniques and procedures. See Table 16.1 for examples.

Hands-on Training

Most institutions have developed procedures for various activities, and the employees who handle infectious and medical wastes should be expected to follow the designated SOPs for the waste-handling activities. Information on these procedures can be presented theoretically through discussion of the risks, alternatives for risk management, the rationale for the SOPs, using the SOPs, and the consequences of not doing so. The material can be presented by lecture, demonstration, pictures and diagrams, and audiovisual materials.

Nevertheless, the best way to learn a procedure is by actual practice. When it is relevant, hands-on training should always be included in the training program.

Table 16.1.   Defining the Course

Approach	Examples
By topic	An overview of the waste management system
	Handling of infectious wastes
	Handling of radioactive wastes
	Handling of chemical wastes
	Regulatory requirements for infectious wastes
	Regulatory requirements for chemical wastes
	Regulatory requirements for radioactive wastes
	Risks of handling infectious wastes
	Managing sharps
	Managing cytotoxic drugs
	Risks of handling chemical wastes
	Risks of handling radioactive wastes and protection from exposure to radioactivity
	Emergency response
By course participants	All waste handlers
	Handlers of infectious wastes
	Handlers of chemical wastes
	Handlers of radioactive wastes
	Those who collect and move the waste
	Treatment equipment operators
	Nurses
	Pharmacists
	Medical technologists
	Laboratory technicians
	Other job classifications
By technique	Discarding sharps and use of sharps containers
	Closing containers of infectious waste
	Collecting and moving waste containers
	Storing radioactive wastes
	Redistilling chemicals
	Operating the steam sterilizer
	Operating the incinerator
	Using personal protective equipment
	Cleaning up a spill

Hands-on training is best done one-on-one or in small groups where careful supervision is possible. Trainees should be able to practice the techniques until they feel comfortable performing them.

The hands-on training part of the course should consist of demonstration as well as practice of correct techniques. It should also include explanation of the reasons for doing something in a particular way, with stress on the importance of doing it the correct way. Question-and-answer sessions are also useful to the course participants.

Testing

Each training course should include testing of employees on the material presented in the course. The testing need not and should not be extensive. Short quizzes with true/false and multiple choice questions are simplest and sufficient.

Such testing serves several important functions. These include:

- identification of employees who need additional training
- identification of problem areas that need emphasis or a different type of presentation in subsequent courses
- evaluation of the effectiveness of the training

The psychological effect of testing is generally beneficial. Knowledge that there will be a test at the end of the course usually provides incentive for learning. Course participants are more likely to have a better attitude toward the course and a positive approach to learning, and they will try to do well on the tests. Therefore, poor test scores usually indicate that the person did not learn the material and should be scheduled for additional courses. It must be remembered, however, that some people always do poorly on tests, despite what they know. This situation can usually be ascertained without too much difficulty; such a person can then be evaluated in an interview or practical test, for example, rather than through formal testing.

Test results can and should be used to evaluate the effectiveness of the individual parts of each course. For example, consistently poor scoring on a certain topic means that that material was not presented effectively. It is important to realize this and to identify the problem so that it will not be perpetuated in subsequent courses. There are various possible solutions to the problem; for example, the material can be restructured to include more or less detail, more time can be devoted to discussing the material, and/or greater emphasis can be placed on the topic.

You want to know that training is effective overall, that the resources (time and money and effort) expended on training activities are worthwhile. Test results are useful for evaluating the general effectiveness of training as well as the individual components. Test scores indicate whether or not course participants understand the material that was presented. Consistently poor test scores reveal that training was not effective. If this happens, the course should be examined for content, approach, topic emphasis, and presentation. Poor training is not only a waste of valuable resources, it may also be counterproductive.

Follow-up to Training

Follow-up to training is an important aspect of the training program. Whereas testing as part of a training course indicates the immediate effectiveness of the training, follow-ups at a later date provide data about retention of information, long-term effectiveness, and the need for refresher courses.

Many options are available for training follow-ups. These include observation of practices (spot checks), written tests, and drills of particular activities such as emergency response techniques (see Chapter 15).

The Training Schedule

Most training programs are not effective if they are limited to a single class. Repeat sessions and refresher courses are essential to promote and restore interest, awareness, and concern.[10]

This need is borne out by several studies done at the time that sharps container systems were introduced to hospitals. These studies demonstrated that the level of awareness about the risks posed by sharps and about the use of special containers for discarded sharps increased during the education "campaign" that accompanied selection of the sharps container system and its introduction to the hospital floors. Interest was initially high, as was compliance with the SOPs for use of the new system. After some time had passed, however, interest dropped off significantly. Interest in the sharps container system was renewed only through a new set of in-service training sessions.

Therefore, it is essential to establish a schedule for presentation of the various types of training courses. A schedule is needed for the initial presentation of the courses as well as for subsequent refresher courses. The schedule should not be rigid. Rather, additional courses should be given as the need arises. For example, spot checks of techniques (see "Follow-up to Training") may indicate the need for refresher courses. This approach can eliminate the need for waiting for an increase in accidents, which would otherwise have to be the indicator that additional training is necessary.

Another advantage of sequential courses is that they serve to inculcate the policies and procedures of the institution. Repetition has its role in training. Additional courses also provide a forum for presenting and explaining any new or changed policies and procedures that have been adopted.

Furthermore, it should be noted that the proposed OSHA rule would require repeat training at least annually.[11]

Employees To Be Trained

Everyone who handles infectious and medical wastes in any way should be included in the training program. For example, "everyone" includes those who actually generate the waste by discarding materials, those who collect and move the waste, those who operate the treatment equipment, and those who handle boxed waste on the loading dock.

Two approaches can be taken to selecting the attendees for any particular training course. The course can be directed toward those with a particular job description, or the course can include all who handle a particular type of waste. As is shown in Table 16.2, each approach has advantages and disadvantages.

The best approach is probably a compromise of sorts. Attendance at the training should be based on waste type. This will allow interaction among all who handle in any way a particular type of waste. They will be able to share concerns about waste that is not handled in accordance with established policies and SOPs. The consequences for others "downstream" will then be apparent because these other persons will be real people rather than a theoretical concept. This should lead to greater concern about following SOPs.

In order to minimize the disadvantages of a class based on waste type, it may be advisable to divide the class into smaller groups for the hands-on part of the training. This will allow for classroom activities that are especially designed for and relevant to the participants of each group.

Table 16.2.   Selecting the Approach to Course Attendance

Approach	Advantages	Disadvantages
By job description	Training is tailored to the specific needs of the group.	No information is provided on what others do before and after this person's role.
		There is no interaction with other handlers of the waste, no opportunity to share concerns.
By waste type	Attendees receive overall picture of waste management.	Effect of training can be dissipated by subjecting employees to irrelevant (for them) information.
	There is opportunity for interaction with others who handle the same type of waste.	

Instructors and Instruction

Qualified instructors are essential for a successful training program. Ideally, the instructor should have experience in teaching training programs and should be familiar with the particular hazards being discussed. Someone with personal experience in handling the wastes being discussed brings a unique perspective and special insights to the training.

Instruction methods are equally important in determining the success of training. It is essential, for example, that the tone of the course not be condescending toward the participants.

Another important issue is the language of instruction and the possible need for multilingual training. A person cannot be trained if he/she cannot understand what is being said. Appropriate questions that address this issue include: Are all the course participants fluent in English? If not, which and how many languages are needed? Is it feasible to offer separate courses for each language, or is it better to present one course with simultaneous translations?

Recordkeeping

It is important to prepare and maintain detailed records on the training program. The records are a regulatory requirement for the right-to-know training program, but they also serve a useful purpose in providing documentation of all training efforts. They provide a database on the who, what, when, and where of the training program.

The following information should be included in the records for the training program:

- a record of the courses given in the training program with schedules for initial and repeat presentations of each course
- a detailed record for each course, including the contents of the course, the location and time of each course, a roster of the attendees with names and job titles, the name and qualifications of the instructor, the schedule for hands-on training, and the tests administered
- a file for course evaluations and a record of all responses made to these evaluations
- a record in each employee's personnel file that includes courses attended, training received, and test results

It is important to have a written record of the training provided to each employee. Such documentation may be essential some day for legal purposes.

REFERENCES

1. *Code of Federal Regulations*, Title 29, Section 1910.1200, "Hazard Communication."
2. U.S. Department of Labor, Occupational Safety and Health Administration. "Chemical Hazard Communication." OSHA Publication #3084 (revised). OSHA, Washington, DC (1987).
3. U.S. Department of Labor, Occupational Safety and Health Administration. "Hazard Communication; Final Rule." *Federal Register* 48(228):53280–53348 (November 25, 1983).

4. U.S. Department of Labor, Occupational Safety and Health Administration. "Hazard Communication; Definition of Trade Secret and Disclosure of Trade Secrets to Employees, Designated Representatives and Nurses; Final Rule." *Federal Register* 51(189):34590–34597 (September 30, 1986).
5. U.S. Department of Labor, Occupational Safety and Health Administration. "Hazard Communication; Final Rule." *Federal Register* 52(163):31852–31886 (August 24, 1987).
6. Joint Commission on Accreditation of Healthcare Organizations. *Accreditation Manual for Hospitals*, 1990 ed. (Chicago: JCAHO, 1989).
7. Standard #PL.6, in *Accreditation Manual for Hospitals*, 1988 and earlier editions.
8. New Jersey Department of Health. Worker and Community Right To Know Act, Occupational Disease Prevention and Information Program. *Guidelines for Education and Training Programs* (August 1985).
9. Gershon, R. M., D. O. Fleming, and D. Vlahov. "Organizing an Effective Training Program to Meet the New OSHA Requirement on HBV/HIV," in *Biohazards Management Handbook,* D. F. Liberman and J. G. Gordon, Eds. (New York: Marcel Dekker, 1989), pp. 417–430.
10. Petty, R. E., and J. T. Cacioppo. *Attitudes and Persuasion: Classic and Contemporary Approaches* (Dubuque, Iowa: William C. Brown Group, 1981).
11. U.S. Department of Labor, Occupational Safety and Health Administration. "Occupational Exposure to Bloodborne Pathogens; Proposed Rule and Notice of Hearing," *Federal Register* 54(102):23042–23139 (May 30, 1989).

CHAPTER 17

Completing the Process: Essential Components of Effective Waste Management

Chapter 1 explains that the objectives of infectious and medical waste management include reducing risks, containing costs, and planning for the future. Cost containment and contingency planning and the alternative of using regional facilities, cooperative agreements, and commercial waste disposal services are discussed in this chapter as additional components to help a medical institution meet those objectives.

TAKING STOCK

This book presents a comprehensive strategy for managing infectious and medical wastes. Because facilities differ, not all components suggested here are appropriate or necessary for every facility to meet its waste management objectives. With the many recent changes in waste management laws and standard practices, few institutions have made satisfactory progress in waste management. An evaluation of the current system of infectious and medical waste management is the starting point for strategic planning.

Audit of Current Practices

Appendix C is a checklist for reviewing current practices with respect to suggestions given in this book. The auditor should be alert for opportunities to reduce risks, costs, and waste generation. A visual inspection allows for some subjective interpretation, so it may be best to include the checklist as part of a more thorough, descriptive status report.

An objective analysis of practices requires gathering quantitative data on waste generation rates and disposal costs for each waste type. Accurate cost accounting puts waste management in perspective with other institutional programs and allows for further analysis for cost containment.

It is also useful to review accident records and to quantify them for incidence and severity. When presenting proposals for improvements to institutional executives, quantitative data are most useful in supporting suggestions for change. Thus, a system of recordkeeping is useful for making objective data analysis.

The identification of trends can be another benefit of quantitative analysis. Spotting a trend can be very helpful for planning, especially when it is supported by regulatory trends, plans of the waste management industry, and trends at other facilities with similar operations.

COST CONTAINMENT

Cost containment may be the most difficult aspect of managing infectious and medical waste. Control of risk and of immediate costs always seem to be at odds with one another. There appears to be no end point for expenditures to achieve an ever smaller level of risk.

The term ''immediate costs'' is used to differentiate current budget costs (short-term) from expenses deferred to a later date (long-term). One purpose of reducing risks is to reduce these potential deferred costs (e.g., workers' compensation claims, site cleanup costs, fines, lawsuits). When risk reduction practices appear to be costly, the business officer must decide if they are justified by savings in potential deferred costs. Be very careful to include long-term costs when applying cost-benefit analysis to this decision, especially when there is the potential of harm to human health; there are many intangibles that are difficult to measure but have real value. The courts have awarded punitive damages when compassion has given way to numerical analysis.

A review of practices, suggested in the previous section, should include a cost analysis of waste management, including time studies that estimate the cost of handling, transport, recordkeeping, and other compliance activities. Information on operational costs, for example, is necessary for making waste management decisions. How much does steam sterilization cost when depreciation, maintenance, and labor are included? For wastes that can be treated by either steam sterilization or incineration, waste treatment decisions should be based (in part) on actual operational costs of each process. Disposal costs are also affected by other decisions (some of which may appear unrelated to waste disposal at the time), such as the decision to use disposable supplies for infection control.

Waste minimization is perhaps the most direct way to control the cost of waste disposal. As discussed in Chapter 10, waste management costs are generally proportional to risks. Since most waste from medical institutions is not infectious, poor separation of infectious and non-infectious waste results in higher overall disposal costs. Thus, source separation is important for cost control. Also, when regulations allow discretion, deciding what wastes are to be considered infectious has a significant effect on costs.[1]

Many waste minimization practices are costly in themselves, however, and require initial and continuing employee training costs. These costs must be considered when deciding upon institutional waste minimization procedures.

Don't attempt cutting the budget for measures that protect against occupational or environmental risk. Procedures and equipment with special features that ensure safety and reduce environmental impact usually are costly but may pay for themselves in lower environmental liability. Don't look to save expenses on commercial waste management services either; the cheapest bid may result in greater liability.

CONTINGENCY PLANNING

Hazardous material emergencies were discussed in Chapter 15, but other crisis situations inevitably occur in the course of waste management. Contingency planning is a necessary part of all enduring and realistic waste management strategies. Make plans for possible failures in the waste management system, including:

- breakdown of the autoclave or incinerator for treating infectious waste, including upset of treatment equipment resulting in its failure to sterilize infectious waste. For example, if an autoclave is used as a backup for an incinerator, routine collection may require special bags that are appropriate for both burning and steam sterilization.
- temporary unavailability of other disposal routes and services for infectious, chemical, or radioactive wastes. There may be periods in which a contracted disposal firm cannot provide services due to a breakdown of its equipment (e.g., truck, incinerator), employee strikes or unavailability of a disposal site.
- failure of waste handling equipment or standard procedures
- breach in security where hazardous materials and wastes are stored and handled, including unexplained loss of inventory

Equipment breakdowns are inevitable. Plan for alternate treatment, disposal, and handling methods.

WORKING WITH COMMERCIAL WASTE MANAGEMENT FIRMS

Hospitals and other medical facilities have traditionally managed infectious and medical waste at their facility (i.e., onsite), using an autoclave or a facility-owned incinerator. This is still true today; only about 15% of infectious waste is treated offsite.[2] However, it is increasingly difficult for hospitals to operate their own incinerators. Required pollution control equipment is affordable only for facilities that serve several generators and can take advantage of economies of scale. For smaller generators of infectious waste, the expense of onsite incineration makes that option even less feasible. Thus, there is a trend toward offsite treatment, especially in those states where infectious waste is more closely regulated. Commercial waste management services are rare in rural areas and in states where infectious and medical waste is loosely regulated.

Waste management firms offer transport, storage, treatment, and disposal services, either by themselves or in conjunction with other facilities. Waste brokers (a type of disposal firm) typically offer transport to a treatment or disposal facility owned by another company. Brokers may offer additional services, such as testing, waste segregation, repackaging, temporary storage, employee training, and compliance assistance. Because they don't have permanent ties to any facility, brokers can collect waste from many generators and seek the least expensive treatment and disposal route. Because the incentives for improper disposal are great, brokers and other transporters can also be a weak link in the control of waste as it moves through the system. Careful selection is important.

The Waste Management Business

Waste management is big business. The two largest disposal firms, Waste Management, Inc. and Browning Ferris Industries, Inc., had 1988 revenues in excess of $3 billion and $2 billion, respectively. Thousands of smaller firms are in waste management also. As an industry, waste management has been one of the fastest-growing sectors of the economy in recent years.

Waste management is a complicated business. Figure 17.1 shows the interrelations between waste generators, transporters, and brokers, as well as insurers and facilities that treat, store, and dispose of waste. Municipal refuse (solid waste) is the primary business of many of these firms, but they increasingly offer specialized services for chemical, infectious, and medical waste. Some firms have built their business on a small niche in the waste disposal market, such as waste solvent recovery or low-level radioactive waste brokerage. As with most other industries, there has been much merger and acquisition activity in the 1980s, and many small firms now are subsidiaries of much larger corporations.

Infectious and medical waste management is only a small part of the entire waste disposal business. Like chemical and other special waste, however, the per-ton disposal cost is higher than for solid waste. In some communities on the east coast, disposal costs for municipal solid waste run $100 per ton, compared with $1000 per ton for infectious wastes.[3] Profit margins also tend to be higher than for normal trash, making infectious and medical waste disposal an attractive business. Whenever government changes infectious and medical waste regulations (or simply

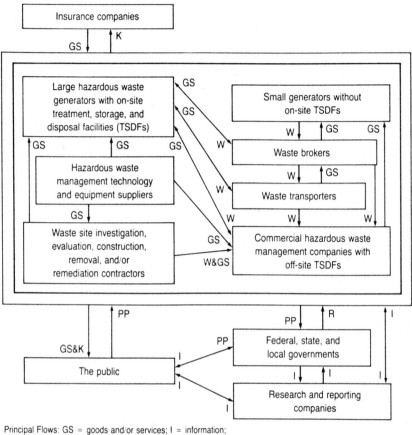

Principal Flows: GS = goods and/or services; I = information;
K = risks; PP = political pressure; R = regulation;
W = hazardous waste

Figure 17.1.   Major interrelations among key parties affecting and affected by hazardous waste management.

changes a definition), significant shifts in the waste disposal market can take place. And these businesses are very much interested in the potential profit in the millions of red bags handled every day.

Risks of Offsite Disposal

Risks to the institution don't end when a contract is signed to send waste off-site to a commercial treatment and disposal facility. The institution may be liable for the safety of employees of firms contracted for waste management, particularly for work that is performed at the institution. As discussed in Chapter 1, an institution is forever liable for its waste and the harm it may cause. If a waste is disposed of improperly, remedial action (e.g., recovery of the waste, cleanup of the site, treatment [if necessary], and disposal) may be the responsibility of the institution. The ease with which unscrupulous operators can set up and stay in business is frightening: the facade of a legitimate waste disposal firm can be created for the cost of renting a truck. Caution is prudent even when dealing with established firms and licensed facilities. There are no guarantees that all currently approved operations pose no risk to human health and the environment.

The decision to use an offsite facility should consider both the short-term costs (treatment and disposal fees) and the potential long-term costs (onsite vs offsite risks). The liability of institutions that contribute wastes to an offsite facility is somewhat reduced because risks are shared by the operator and other generators that contribute wastes. New risks include transportation risks (e.g., an accident resulting in a spill), the lack of direct control of the disposal firm's operation (i.e., the institution is vulnerable to disruptions in service) and the possibility of environmental liability. Because of this, it is very important that commercial treatment and disposal facilities be inspected and evaluated, both prior to sending waste there and during the term of the contract.

Evaluating Offsite Facilities

The objective of facility evaluations is to ensure that the waste is being managed properly, that the potential for harm to human health and the environment is as small as possible. Every step and every firm used for offsite waste management should be thoroughly investigated: transporter, broker, treatment facility, and landfill. With such a broad scope, the evaluation of a waste management firm, its facility, and its potential impacts can become a complex and time-consuming endeavor.[4]

Facility evaluations can begin with a series of questions to the commercial firms; some examples follow:

Occupational and Environmental Safety

- What precautions are used to prevent a release to the environment?
- Have tests been conducted to measure releases to air, water or sanitary sewer? Has incinerator ash been tested for toxic metals and products of combustion? What are the results?
- Do you have routine, permitted releases?
- Have you ever had an unplanned release to the environment?
- What precautions do you take to prevent direct contact of hazardous materials and wastes to the public, or routine releases?
- What is the fate of the treated waste? How is incinerator ash disposed of?

- How are your employees trained to safely handle infectious and medical wastes, and comply with occupational safety and environmental protection rules?
- Have you ever been fined for violations of occupational safety and environmental protection laws (or received a citation or letter of warning)?

Other Liabilities

- How many other generators use your services? What kinds of businesses are represented?
- How do you prevent other generators from placing inappropriate wastes into your collection containers?
- Do you handle any other hazardous waste or materials?
- How are records kept to differentiate among waste generators, so that careless generators can be identified?
- How long have you been in the waste disposal business? What is your business plan? (Review the company's annual report for an indication of financial strength and financial liabilities.)
- What insurance and forms of financial assurance mechanisms do you have? (Have the institutional risk manager review this response.)
- Are you involved in any compliance orders or lawsuits?
- What TSD facilities do you plan to use? Do you own them? If not, what assurances do you have of future access?

Emergency and Contingency Planning

- Are there plans for hazardous material emergencies (e.g., fire, spill, or other release)?
- What contingency plans are there for an extended breakdown of your vehicles, storage facilities (e.g., refrigerators), or treatment equipment?

Additional Services

- Will you provide training to the institutional staff responsible for waste management?
- What assistance will you offer for our compliance and risk management efforts?

Federal, state, and local regulators should also be queried to confirm the status of licenses and to learn if there have been any compliance actions against the firm. Other references to be checked are other generators that use the site (they may have conducted their own evaluation—and may be willing to share the results), trade associations, and local environmental organizations that may be concerned about the facility's operation.

A site visit of the facilities by a member of the institution is likely to be very valuable; the authors highly recommend it. The general appearance of a treatment, storage, or disposal facility is a good indicator of its overall concern for good operation and protecting the environment. Note, however, that visual observations can't measure treatment efficiency or stack emissions. Be especially careful when reviewing a facility that also manages other types of waste. The potential for environmental harm from chemical waste is much greater than that from infectious and medical waste. Inadequate recordkeeping by the firm could make you responsible for all wastes handled at the facility.

Most waste disposal firms welcome inspections of their facilities by their customers. Be wary if the firm is reluctant to allow a visit, a review of its records, or photographs. Once on site, ask a lot of questions; talk to workers who are not part of the "tour." A checklist may be useful.[4] Commercial storage, treatment, and disposal facilities should have all of the components of a waste management system that are described in this book: a waste management plan, an occupational safety program, contingency and emergency plans, and an employee training program. In short, the contractor's efforts to reduce environmental and occupational risk should be just as thorough as the generator's.

Take time—it can take several hours—to thoroughly review the facility's records. Examine the licenses, permits, and any special conditions that may be part of them. Make sure that records of waste received and treated are kept orderly, and that ownership and disposition of the waste is clearly identified: institutional liability is highly dependent on this.

If travel to a site is too costly due to distance, consider joining with other generators—or calling nearby generators—that use the site to perform an evaluation.

Bid Specifications and Requests for Proposals

Many institutions obtain the services of a waste disposal firm through a competitive bid process. As has been suggested in this chapter, the bidder with the lowest disposal fees may put the institution at risk for long-term liabilities. Bids need to include conditions that protect the institution, most importantly:

- disclosure of subcontractors who may provide treatment, disposal, or transport services
- receipt of copies of manifests (shipping documents, including those for remanifested waste) and any other records pertaining to the disposition of the waste
- notification of any changes in insurance coverage, subcontractors, transporter, or treatment or disposal sites; any releases; or any citations for noncompliance
- certification of waste destruction or disposal

Other conditions that can be included in the bid specifications are inferred in evaluation questions listed under "Evaluating Offsite Facilities." Ask a large generator in the area for a copy of their bid specifications to use as a model.

An alternate method of competitively seeking waste disposal services is through the use of a Request for Proposal (RFP). The RFP process is a performance-based purchasing process, whereas competitive bidding seeks the lowest (short-term) disposal fees. An RFP describes the problem (stating the amount of waste needing disposal and other waste management needs) and the general criteria under which the services are to be performed. Price is usually a minor component of the RFP evaluation criteria (less than 30%). Noncost criteria include low environmental impact, safety measures, proposed procedures for managing waste, and other special needs of the institution.

The RFP process is desirable because the burden of describing risk management methods lies with the proposers. Also, because disposal firms don't have to conform to detailed bid conditions, an RFP opens the process to unique or new solutions. A proposer, for example, might offer a new treatment technology that promises low cost and risk reduction. However, RFPs are much more difficult to evaluate than bids and require a method to judge environmental impact.

Bids can contain some aspects of an RFP. For example, a clause can be added that allows bidders to offer alternate technologies if other conditions are met. Written this way, bids become detailed, lengthy documents. For small waste volumes, potential vendors may resist participating in either process—they perceive that the cost of preparing a response is not worth the effort. Firms may need to be encouraged to submit bids and proposals. A discussion with vendors may help you decide on the best method of seeking such services.

Whatever method is used to obtain a disposal contract, it should first be approved by the institution's attorney. The risk manager can evaluate insurance coverage. Remember that a contract is usually no defense against legal liabilities, even for contracts that have "hold harmless" or indemnification wording. As explained in Chapter 1, liability may be shared with haulers, treatment facilities, and disposal sites, but the generator always retains some liability for the waste. In some cases, that liability may be disproportionate to the generator's role in the waste management system, the degree of hazard, or amount of waste relative to other wastes handled by the contracted firms.

Resources

Trade groups to which the institution belongs may be helpful in finding and selecting disposal vendors. State industry associations should be active in protecting the institution from laws that favor disposal firms over waste generators when deciding liability and responsibility for requirements. EPA's Small Business Ombudsman (1-800-368-5888) may be able to assist if there is evidence of monopolistic practices.

REGIONAL FACILITIES AND COOPERATIVE ARRANGEMENTS

Another offsite solution to managing medical waste is to join with other generators in sharing the use of a facility. Two types of shared waste disposal facilities have evolved: facilities that are owned by a hospital (or other institution) and agree to accept wastes from other generators, and facilities that are owned jointly by several hospitals in a partnership or cooperative business. The cooperative venture typically operates as a nonprofit business and may choose to accept wastes from nonmember institutions and small quantity generators. Other aspects of this alternative are discussed in Chapter 5.

Risks of Joint Arrangements

Accepting waste from area generators (e.g., clinics, doctor's offices, nursing homes, other hospitals) for treatment at the institution also subjects the institution to the liabilities and controversies of the waste disposal business. Although such activity can reap short-term profits, they can easily be offset if environmental harm results. Some hospitals provide disposal services to small medical waste generators as a valuable and necessary community service. In rural areas, small generators have little choice but to ask for help. Still, when deciding to accept offsite waste, institutions need to carefully decide whether it is worth the potential risk.

Joint ownership of a facility with other waste generators may actually increase an institution's risk. If one hospital fails to separate chemical and infectious waste before sending it to a jointly used infectious waste incinerator, all participants could be liable for illegally treating chemical waste. Suddenly, all of the partners in the cooperative are in the waste disposal business, incurring all the commensurate liabilities.

DON'T STOP HERE

This book attempts to explain a very detailed subject with the goal of helping generators of infectious and medical waste. The objective is not only to reduce costs, but to control the risks of waste management. The authors encourage continued efforts to keep up with changes in this field. The references given throughout the book offer further insights into this complex area.

When implementing a waste management program, it is important to keep the problem in perspective. Waste affects many people, and their concerns are not to be brushed aside. But all institutions conduct other activities that also impact the environment and people and many of these impacts are more important than those caused by infectious and medical waste. The authors hope that this book will help waste generators understand and reduce risks while formulating a rational waste management strategy.

REFERENCES

1. Rutala, W. "Cost-Effective Application of the Centers for Disease Control Guidelines for Handwashing and Hospital Environmental Control," *American Journal of Infectious Control* 13:218–224 (1984).
2. Lee, C. C., et. al. "A Review of Biomedical Waste Disposal: Incineration," U.S. Environmental Protection Agency, Office of Research and Development (February 19, 1988).
3. Peters, R. A. "Future Trends in Infectious Waste Management," paper presented at Evaluating Your Infectious Waste Disposal Contractors course, Department of Engineering Professional Development, University of Wisconsin–Madison (December 16, 1988).
4. Goldman, B. A., J. A. Hulme, and C. Johnson. *Hazardous Waste Management: Reducing The Risk* (Washington, DC: Island Press, 1986).

APPENDICES

A Guide to the Medical Waste Tracking Regulations

WHAT IS THE PURPOSE OF THE MEDICAL WASTE TRACKING REGULATIONS?

Congress passed the Medical Waste Tracking Act of 1988 because of its concern about beach washups of infectious waste. This statute directed EPA to develop regulations to track medical wastes as part of its RCRA program. EPA promulgated medical waste tracking regulations in March 1989.[1] These regulations set up a two-year demonstration program that is intended to assess the effectiveness of regulations in eliminating improper disposal of medical wastes. The regulations apply to regulated medical wastes generated in certain "Covered States" during the period from June 22, 1989 to June 22, 1991.

WHICH WASTES ARE REGULATED?

The medical waste tracking regulations pertain to "regulated medical wastes." A regulated medical waste is defined as a solid waste that is generated in the diagnosis, treatment, or immunization of human beings or animals, in related research, or in the production or testing of biologicals.[2,3] In addition, the waste must be included in one of the seven listed classes of regulated medical wastes. These classes are:

- cultures and stocks
- pathological wastes

- human blood and blood products
- sharps
- animal waste
- isolation wastes
- unused sharps

EPA used three criteria for listing and defining these classes: actual or potential infectiousness, physical hazard, and potential aesthetic degradation of the environment. As a result, some classes are broader than their names imply, and some include wastes that are only marginally related to the class name. Therefore, it is important to check the definitions. See Table A.1 for the types of waste that are included in each class of regulated medical waste.

Table A.1.   Classes of Regulated Medical Wastes[2,3]

Class	Description
Cultures and Stocks	Cultures and stocks of infectious agents. Includes: • Cultures from medical and pathological laboratories • Cultures and stocks of infectious agents from research and industrial laboratories • Wastes from the production of biologicals • Discarded live and attenuated vaccines • Culture dishes and devices used to transfer, inoculate, and mix cultures
Pathological Wastes	Human pathological wastes. Includes: • Tissues, organs, and body parts and body fluids that are removed during autopsy or surgery or other medical procedures • Specimens of body fluids and their containers
Human Blood and Blood Products	• Liquid waste human blood • Products of blood • Items saturated and/or dripping with human blood • Items that were saturated and/or dripping with human blood that are now caked with dried human blood Includes: • Serum, plasma, and other blood components, and their containers, which were used or intended for use in either patient care, testing, and laboratory analysis or the development of pharmaceuticals • Intravenous bags
Sharps	Sharps that have been used in animal or human patient care or treatment or in medical, research, or industrial laboratories. Includes: • Hypodermic needles • Syringes (with or without the attached needle)

Table A.1, continued

Class	Description
	• Pasteur pipettes • Scalpel blades • Blood vials • Needles with attached tubing • Culture dishes (regardless of presence of infectious agents) • Other types of broken or unbroken glassware that were in contact with infectious agents, such as used slides and cover slips
Animal Waste	Contaminated animal carcasses, body parts, and bedding of animals that were known to have been exposed to infectious agents during research (including research in veterinary hospitals), production of biologicals, or testing of pharmaceuticals
Isolation Wastes	• Biological waste and discarded materials contaminated with blood, excretion, exudates, or secretions from humans who are isolated to protect others from certain highly communicable diseases • Such materials from isolated animals known to be infected with highly communicable diseases
Unused Sharps	The following unused, discarded sharps: • Hypodermic needles • Suture needles • Syringes • Scalpel blades

The regulations specifically exclude from regulation as regulated medical waste five types of waste:

- hazardous waste identified or listed in 40 CFR 261
- household waste as defined in 40 CFR 261.4(b)(1)
- ash from the incineration of regulated medical waste
- residues from processes that both treat and destroy the waste
- human corpses, remains, and anatomical parts that are intended for interment or cremation

The regulations also exempt from regulation as regulated medical waste two types of waste:

- etiological agents being transported interstate in accordance with all applicable shipping regulations
- samples of regulated medical waste transported offsite by EPA- or state-designated enforcement personnel for enforcement purposes

There are also special provisions for waste mixtures that include regulated medical waste. Specifically:

- A mixture of solid waste and regulated medical waste is considered a regulated medical waste.
- Mixtures of hazardous waste (per 40 CFR 261) and regulated medical waste are subject to the medical waste requirements unless the hazardous waste manifest must be used (40 CFR 262 or 266).

IN WHICH STATES ARE THE REGULATIONS IN EFFECT?

The medical waste tracking regulations do not apply nationwide. Rather, they establish a demonstration program for certain "Covered States." These states are Connecticut, New Jersey, New York, Rhode Island, and Puerto Rico.[4] All regulated medical waste generated in these states is regulated. Such waste is regulated even if it is transported out of the Covered State where it was generated into a non-Covered State. Waste brought into a Covered State is subject to regulation unless it is proven that the waste was generated in a non-Covered State.

WHAT IS MEDICAL WASTE TRACKING?

A tracking form must be used when regulated medical waste is transported offsite from the facility where it was generated. The multiple-copy tracking form identifies the generator, transporter, and destination facility and lists the amount of waste (number of containers and total weight or volume) and type of waste (untreated or treated). The tracking form must accompany each waste shipment from the generator to the point of treatment and destruction* or to the point of disposal. A copy of the tracking form must be returned to the generator to confirm that the shipment reached its intended destination.

Signatures are required on the tracking form each time the waste changes hands. The generator's signature certifies that the information on the form is correct. Each transporter, intermediate handler, and destination facility signs to verify receipt of the waste specified on the form. Therefore, the tracking form provides a record of the chain of custody for each waste shipment.

---

*When destroyed, the waste is no longer generally recognizable as medical waste. Destruction processes include incineration, grinding, shredding, crushing, and melting.

All copies of the tracking form and other records must be maintained for three years. In addition, the required reporting of discrepancies* and exceptions† should allow the tracing of lost or mismanaged shipments of regulated medical waste.

WHAT IS THE SCOPE OF THE MEDICAL WASTE TRACKING REGULATIONS?

The regulations require tracking each shipment of regulated medical waste from the generator to the point of treatment-and-destruction or disposal. Other aspects of the regulations are designed to provide protection to waste handlers and the public by specifying types of containers and vehicles that must be used for transport of regulated medical wastes.

The regulations include requirements that pertain to:

- handling of the waste before transport
- generators of regulated medical waste
- transporters of regulated medical waste
- treatment, destruction, and disposal facilities

WHAT ARE THE PRETRANSPORT REQUIREMENTS?

The pretransport requirements for regulated medical waste pertain to:

- waste segregation
- packaging of the waste
- storage of the waste
- decontamination of reusable containers
- labeling of containers
- marking of containers

Segregation Requirements

Regulated medical waste must be segregated from all other waste generated at the facility, as much as practicable. In addition, fluids (defined as quantities

---

*Discrepancies are (1) a difference between the number of containers or the type of waste received and that listed on the tracking form; (2) packaging that is broken, torn, or leaking; and (3) a missing, incomplete, or unsigned tracking form.

†An exception means that the generator has not received a copy of the tracking form from the destination facility within 35 days after the waste was shipped offsite.

in excess of 20 cc) and sharps are to be segregated from other types of regulated medical waste (40 CFR 259.40).

Packaging Requirements

The regulations list specifications for containers that are used to transport regulated medical waste (40 CFR 259.41). They must be rigid, leak-resistant, impervious to moisture, and of sufficient strength to prevent tearing and bursting under normal use and handling. Each container must be sealed to prevent leakage during transport. In addition, sharps containers must be puncture resistant, and containers for fluids must be break resistant with tight caps or stoppers. Oversized medical waste does not have to be placed in a container.

Storage

The regulations specify storage conditions at a central collection point for regulated medical waste before it is treated or disposed of onsite or transported offsite away from the facility where it was generated (40 CFR 259.42). Integrity of the packaging must be maintained during storage, and the waste must be protected from water, rain, and wind. The waste must be maintained in a nonputrescent state so refrigeration may be necessary. Storage areas must be kept locked, and access must be limited to authorized personnel. The storage area must be kept free of animals and pests.

Decontamination of Reusable Containers

Regulated medical waste may be transported in reusable containers if they are rigid and clean. EPA suggests that liners be used and that the containers be washed and decontaminated before reuse. Liners for containers may not be reused; they must be managed as regulated medical waste. Containers must be decontaminated before reuse if they are visibly contaminated; if they cannot be cleaned, they must be managed as regulated medical waste (40 CFR 259.43).

Labeling Requirements

Labels are required to identify containers of untreated medical waste. Four alternatives are acceptable for labeling. These are (1) the wording "Medical Waste," (2) the wording "Infectious Waste," (3) display of the universal biohazard symbol, and (4) use of red containers. The labels may be affixed to or printed directly on the container; they must be waterproof (40 CFR 259.44).

There are no labeling requirements for containers of treated medical waste; however, they must be marked (see below).

Marking Requirements

Markings are required on each container to identify the generator and transporter by name and permit or identification number (or address) (40 CFR 259.45). The markings must also include the date of shipment and identify the contents as medical waste. Inner containers must also be marked with the name and number (or address) of the generator.

WHAT ARE THE REQUIREMENTS FOR GENERATORS OF REGULATED MEDICAL WASTE?

The regulatory requirements for generators of regulated medical waste differ according to the quantity of waste generated and what is done with it. There are six possibilities:

- More than 50 lb of regulated medical waste is generated and transported offsite per month.
- Less than 50 lb of regulated medical waste is generated and transported offsite per month.
- Waste is shipped between a generator's facilities.
- Regulated medical waste is incinerated onsite.
- Waste is treated and destroyed or disposed of onsite.
- Waste is exported.

A generator is subject to all applicable requirements.

More Than 50 Lb/Month Transported Offsite

Generators who ship offsite more than 50 lb of regulated medical waste per month are subject to pretransport, tracking form, recordkeeping, and reporting requirements for these wastes. Also, they must ship their waste only with transporters who have notified EPA of their intent to transport regulated medical waste. (See Sections 259.50(e)(1), 259.52, 259.54(a), 259.55, and 259.50(f) of CFR Title 40, respectively.)

These generators must comply with all the pretransport requirements that are detailed above. These are the requirements for segregation, packaging, storage, labeling, and marking.

The generator must originate the tracking form, making enough copies for himself, the transporter(s), any intermediate handlers, and the destination facility (two copies, one to be returned to the generator). In signing the form, the generator certifies that the information is complete and correct. Criminal penalties may be assessed when material information is intentionally omitted or falsified.

Recordkeeping requirements are for three years. Copies of the tracking forms for each waste shipment must be retained as well as copies of any exception reports.

Exception reports must be made and submitted to the state and to the EPA Regional Administrator. They must document efforts to locate the waste and must include a copy of the original tracking form.

Less Than 50 Lb/Month Transported Offsite

When less than 50 lb of regulated medical waste is generated and shipped off-site in any month, regulatory requirements are reduced. (See 40 CFR 259.51(a).) These generators are exempt from the requirements to use a transporter who has notified EPA, to use the tracking form, and to meet the vehicle requirements, provided that certain conditions are met. That is, the waste must be transported by the generator from the point of generation (1) to a health care facility, an intermediate handler, or a destination facility with whom there is a written agreement to accept the waste, or (2) to the generator's place of business.

There are recordkeeping requirements if the generator transports the waste himself or uses a transporter who has notified EPA. (See 40 CFR 259.53(b).) The generator must keep a log of all waste shipments, those the generator transports himself as well those given to a transporter.

Waste Shipped Between a Generator's Facilities

When regulated medical waste is transported from the generation point to a central collection point, the generator is exempt from the requirements to use the tracking form, to use a transporter who has notified EPA, and to meet the vehicle requirements provided that certain conditions are met. (See 40 CFR 259.51(b).) These are (1) the waste must be transported by the generator to a central collection point or treatment facility owned or operated by the generator, and (2) the original generation point and the central collection point or treatment facility must be located in the same Covered State.

These generators must maintain shipment logs at both the original generation point and the central collection point. (See 40 CFR 259.54(a)(2).)

Onsite Incineration

Generators of regulated medical waste who incinerate the waste onsite must keep an operating log of the incinerator operations (40 CFR 259.61). Required information includes the starting date and length of each incineration cycle, the total quantity of medical waste incinerated, and the quantity of regulated medical waste incinerated during each cycle.

When regulated medical waste from other generators is accepted for incineration, records must be kept identifying the generator and the amount of regulated medical waste in each shipment.

Two reports on incinerator operations must be submitted to EPA, one for the six-month period ending December 22, 1989, the other for the six-month period ending December 22, 1990 (40 CFR 259.62).

The receipt of waste from other generators accompanied by a tracking form triggers requirements for processing the tracking form, resolving/reporting discrepancies, and recordkeeping. (See Sections 259.81(a), 259.82, and 259.83 of CFR Title 40.)

Other Onsite Treatment and Destruction or Disposal

When a generator treats and destroys or disposes of regulated medical waste onsite by methods other than incineration, he is not subject to the tracking requirements for this waste, but must keep certain records. (See Sections 259.50(c) and 259.54(c) of CFR Title 40.) The required information includes the quantities of regulated medical waste and total waste treated and destroyed as well as information on waste received from other generators who need not use the tracking form (identification of the generator, weight of waste received, and dates the waste was received and treated and destroyed).

Exported Waste

When regulated medical waste is exported for treatment and destruction or disposal, the generator must request that the destination facility provide written confirmation that the waste was received. If such documentation is not received within the prescribed period, the generator must file an exception report (40 CFR 259.53).

WHAT ARE THE TRANSPORTER REQUIREMENTS?

The regulations give transporters of regulated medical waste a key role in the demonstration tracking program. Therefore, in addition to the expected requirements for vehicles and storage, there are special requirements for waste tracking (notification, use of the tracking form, recordkeeping, and reporting).

Notification

Each transporter must notify EPA and the Covered State(s) of intent to transport regulated medical waste generated in a Covered State before accepting such waste for transport. EPA issues an EPA Medical Waste Identification Number for each Covered State in which the transporter will be operating. This number must be used on the tracking form.

Use of the Tracking Form

The transporter must verify that containers of regulated medical waste are properly labeled and marked and that the information on the tracking form (number and weight of containers) is correct. He must deliver all the waste to the party listed on the tracking form and obtain the generator's permission for an alternate destination (40 CFR 259.74 and 259.75).

Regulated medical waste may be accepted without a tracking form only from generators who generate and ship less than 50 lb of waste per month. The transporter is required to sign the generator's log and to keep a log of these shipments (Section 259.74(g) of CFR Title 40). The transporter must prepare a tracking form for waste from these generators.

The transporter may consolidate shipments of less than 220 lb and remanifest them to a single tracking form. He is then responsible for sending back to the generators copies of the form signed by the destination facility (40 CFR 259.76).

Vehicle Requirements

The regulations specify requirements for vehicles that are used to transport regulated medical waste (40 CFR 259.73). The vehicle must have a fully enclosed, leak-resistant cargo-carrying body. It must be maintained in good sanitary condition. It must be locked if left unattended. The transporter must be identified on the sides and back of the vehicle, and a sign must identify the contents as "medical waste" or "regulated medical waste."

There are also some procedural requirements. The waste must not be subjected to mechanical stress or compaction during loading, unloading, or transport. Mixed waste in the same container must all be managed as regulated medical waste.

Recordkeeping

There are recordkeeping requirements for tracking forms, shipment logs, consolidation logs, and reports (40 CFR 259.77).

Reporting

Transporters of regulated medical waste must submit reports on the source and disposition of the transported waste (40 CFR 259.78). There are four reporting periods (each six-month period during the two years of the demonstration program), and a separate report is required for each Covered State for each reporting period. The reports must provide information on the transporter, number of generators serviced, the identity of and amount of waste accepted from each generator, amount of waste delivered to an intermediate handler or destination facility, and amount of waste delivered to another transporter or transfer facility. If waste is delivered to an intermediate handler or destination facility, the transporter must also identify each such party and report the amount of waste delivered.

WHAT ARE THE REQUIREMENTS FOR TREATMENT, DESTRUCTION, AND DISPOSAL FACILITIES?

These facilities are of three types:

- destination facilities where the waste is treated and destroyed
- intermediate handlers where the waste is either treated or destroyed (but not both)
- disposal facilities where waste that does not meet the treated-and-destroyed criterion is disposed of

Use of the Tracking Form

All of these facilities must process the tracking form. Destination facilities and disposal facilities must return signed copies of the tracking form to the generators acknowledging receipt of the waste. Intermediate handlers must sign the tracking form to acknowledge receipt of the waste; after the waste is treated, they must initiate a new tracking form for shipment of the waste to a disposal facility (40 CFR 259.81).

Recordkeeping

Copies of tracking forms must be maintained for three years. This applies to all tracking forms, those received with incoming waste as well as those initiated for shipments of waste from intermediate handlers. Intermediate handlers must also maintain logs matching incoming and outgoing shipments. Copies of discrepancy reports must also be maintained (40 CFR 259.83).

Reporting

Destination facilities, intermediate handlers, and disposal facilities must try to resolve any discrepancies. Discrepancies that cannot be resolved must be reported (40 CFR 259.82).

WHAT WILL HAPPEN AFTER THE DEMONSTRATION PROGRAM?

The medical waste tracking demonstration program will last until June 22, 1991. From the reports submitted, EPA will establish a baseline of data on medical waste disposal practices. EPA must then submit to Congress a report on medical wastes and on the demonstration tracking program. The report must include:

- an analysis of medical wastes generators, types and quantities of waste generated, and management methods
- an evaluation of the threat to human health and the environment posed by medical waste and its incineration
- estimates of the costs associated with the tracking and management requirements of the demonstration program
- an evaluation of the success of the program and resulting changes in incineration and storage practices
- an examination of methods for managing and treating medical wastes, including the factors influencing the effectiveness of treatment methods
- an evaluation of the appropriateness of the Act's provision for penalties for ensuring compliance
- an evaluation of the effect of excluding households and small quantity generators from regulation

Further action by Congress will be affected by EPA's report to Congress on the demonstration tracking program, as well as by ATSDR's report on the public health implications of medical waste. There are several possible alternatives:

- continuation of the medical waste tracking demonstration program
- termination of medical waste tracking regulations
- retention of regulations in the Covered States
- expansion of the medical waste tracking regulations to the entire nation
- transfer of responsibility for regulation of medical wastes to the states

It is impossible to predict which course of action Congress will select.

REFERENCES

1. U.S. Environmental Protection Agency. "Standards for the Tracking and Management of Medical Waste; Interim Final Rule and Request for Comments." *Federal Register* 54(56):12326–12395 (March 24, 1989).
2. U.S. Environmental Protection Agency. "Standards for the Tracking and Management of Medical Waste; Interim Final Rule and Request for Comments." *Federal Register* 54(56):12373–12374 (March 24, 1989).

3. *Code of Federal Regulations* Title 40, Section 259.30(a), ''Definition of Regulated Medical Waste.''
4. U.S. Environmental Protection Agency. ''Standards for the Tracking and Management of Medical Waste. Amendment to Final Rule.'' *Federal Register* 54(163):35189–35191 (August 24, 1989).

APPENDIX B

U.S. EPA and State Hazardous Waste Contacts

RCRA/Superfund Hotline 1-800-424-9346 (In Washington, D.C.:382-3000)	EPA Small Business Ombudsman Hotline 1-800-368-5888 (In Washington, D.C.:557-1938)	National Response Center 1-800-424-8802 (In Washington, D.C.:426-2675)

Regions	Regions	Regions	Regions
4 — Alabama	5 — Indiana	9 — Nevada	4 — Tennessee
10 — Alaska	7 — Iowa	1 — New Hampshire	6 — Texas
9 — Arizona	7 — Kansas	2 — New Jersey	8 — Utah
6 — Arkansas	4 — Kentucky	6 — New Mexico	1 — Vermont
9 — California	6 — Louisiana	2 — New York	3 — Virginia
8 — Colorado	1 — Maine	4 — North Carolina	10 — Washington
1 — Connecticut	3 — Maryland	8 — North Dakota	3 — West Virginia
3 — Delaware	1 — Massachusetts	5 — Ohio	5 — Wisconsin
3 — D.C.	5 — Michigan	6 — Oklahoma	8 — Wyoming
4 — Florida	5 — Minnesota	10 — Oregon	9 — American Samoa
4 — Georgia	4 — Mississippi	3 — Pennsylvania	9 — Guam
9 — Hawaii	7 — Missouri	1 — Rhode Island	2 — Puerto Rico
10 — Idaho	8 — Montana	4 — South Carolina	2 — Virgin Islands
5 — Illinois	7 — Nebraska	8 — South Dakota	

U.S. EPA REGIONAL OFFICES

EPA Region I
State Waste Programs Branch
JFK Federal Building
Boston, Massachusetts 02203
(617) 223-3468
Connecticut, Massachusetts, Maine,
New Hampshire, Rhode Island, Vermont

EPA Region II
Air and Waste Management Division
26 Federal Plaza
New York, New York 10278
(212) 264-5175
New Jersey, New York, Puerto Rico,
Virgin Islands

EPA Region III
Waste Management Branch
841 Chestnut Street
Philadelphia, Pennsylvania 19107
(215) 597-9336
Delaware, Maryland, Pennsylvania,
Virginia, West Virginia,
District of Columbia

EPA Region IV
Hazardous Waste Management Division
345 Courtland Street, N.E.
Atlanta, Georgia 30365
(404) 347-3016
Alabama, Florida, Georgia,
Kentucky, Mississippi, North
Carolina, South Carolina, Tennessee

EPA Region V
RCRA Activities
230 South Dearborn Street
Chicago, Illinois 60604
(312) 353-2000
Illinois, Indiana, Michigan,
Minnesota, Ohio, Wisconsin

EPA Region VI
Air and Hazardous Materials Division
1201 Elm Street
Dallas, Texas 75270
(214) 767-2600
Arkansas, Louisiana, New Mexico,
Oklahoma, Texas

EPA Region VII
RCRA Branch
726 Minnesota Avenue
Kansas City, Kansas 66101
(913) 236-2800
Iowa, Kansas, Missouri, Nebraska

EPA Region VIII
Waste Management Division (8HWM-ON)
One Denver Place
999 18th Street, Suite 1300
Denver, Colorado 80202-2413
(303) 293-1502
Colorado, Montana, North Dakota,
South Dakota, Utah, Wyoming

EPA Region IX
Toxics and Waste Management Division
215 Fremont Street
San Francisco, California 94105
(415) 974-7472
Arizona, California, Hawaii,
Nevada, American Samoa, Guam,
Trust Territories of the Pacific

EPA Region X
Waste Management Branch—MS-530
1200 Sixth Avenue
Seattle, Washington 98101
(206) 442-2777
Alaska, Idaho, Oregon, Washington

STATE HAZARDOUS WASTE MANAGEMENT AGENCIES

ALABAMA
Alabama Department of
  Environmental Management
Land Division
1751 Federal Drive
Montgomery, Alabama 36130
(205) 271-7730

ALASKA
Department of Environmental
  Conservation
P.O. Box 0
Juneau, Alaska 99811
Program Manager: (907) 465-2666
Northern Regional Office
  (Fairbanks): (907) 452-1714
South-Central Regional Office
  (Anchorage): (907) 274-2533
Southeast Regional Office
  (Juneau): (907) 789-3151

AMERICAN SAMOA
Environmental Quality Commission
Government of American Samoa
Pago Pago, American Samoa 96799
Overseas Operator
(Commercial Call (684) 663-4116)

ARIZONA
Arizona Department of
  Health Services
Office of Waste and Water Quality
2005 North Central Avenue
  Room 304
Phoenix, Arizona 85004
Hazardous Waste Management:
  (602) 255-2211

ARKANSAS
Department of Pollution Control
  and Ecology
Hazardous Waste Division
P.O. Box 9583
8001 National Drive
Little Rock, Arkansas 72219
(501) 562-7444

CALIFORNIA
Department of Health Services
Toxic Substances Control Division
714 P Street, Room 1253
Sacramento, California 95814
(916) 324-1826

State Water Resources Control Board
Division of Water Quality
P.O. Box 100
Sacramento, California 95801
(916) 322-2867

COLORADO
Colorado Department of Health
Waste Management Division
4210 E. 11th Avenue
Denver, Colorado 80220
(303) 320-8333 Ext. 4364

CONNECTICUT
Department of Environmental
  Protection
Hazardous Waste Management
  Section
State Office Building
165 Capitol Avenue
Hartford, Connecticut 06106
(203) 566-8843, 8844

Connecticut Resource Recovery
  Authority
179 Allyn Street, Suite 603
Professional Building
Hartford, Connecticut 06103
(203) 549-6390

DELAWARE
Department of Natural Resources
  and Environmental Control
Waste Management Section
P.O. Box 1401
Dover, Delaware 19903
(302) 736-4781

DISTRICT OF COLUMBIA
Department of Consumer and
  Regulatory Affairs
Pesticides and Hazardous Waste
  Materials Division
Room 114
5010 Overlook Avenue, S.W.
Washington, D.C. 20032
(202) 767-8414

FLORIDA
Department of Environmental
  Regulation
Solid and Hazardous Waste Section
Twin Towers Office Building
2600 Blair Stone Road
Tallahassee, Florida 32301
RE: SQG's
(904) 488-0300

GEORGIA
Georgia Environmental Protection
  Division
Hazardous Waste Management
  Program
Land Protection Branch
Floyd Towers East, Suite 1154
205 Butler Street, S.E.
Atlanta, Georgia 30334
(404) 656-2833
Toll Free: (800) 334-2373

GUAM
Guam Environmental Protection
  Agency
P.O. Box 2999
Agana, Guam 96910
Overseas Operator
(Commercial Call (671) 646-7579)

HAWAII
Department of Health
Environmental Health Division
P.O. Box 3378
Honolulu, Hawaii 96801
(808) 548-4383

IDAHO
Department of Health and Welfare
Bureau of Hazardous Materials
450 West State Street
Boise, Idaho 83720
(208) 334-5879

ILLINOIS
Environmental Protection Agency
Division of Land Pollution Control
2200 Churchill Road, #24
Springfield, Illinois 62706
(217) 782-6761

INDIANA
Department of Environmental
  Management
Office of Solid and Hazardous Waste
105 South Meridian
Indianapolis, Indiana 46225
(317) 232-4535

IOWA
U.S. EPA Region VII
Hazardous Materials Branch
726 Minnesota Avenue
Kansas City, Kansas 66101
(913) 236-2888
Iowa RCRA Toll Free:
  (800) 223-0425

KANSAS
Department of Health and
  Environment
Bureau of Waste Management
Forbes Field, Building 321
Topeka, Kansas 66620
(913) 862-9360 Ext. 292

KENTUCKY
Natural Resources and
  Environmental Protection Cabinet
Division of Waste Management
18 Reilly Road
Frankfort, Kentucky 40601
(502) 564-6716

LOUISIANA
Department of Environmental
  Quality
Hazardous Waste Division
P.O. Box 44307
Baton Rouge, Louisiana 70804
(504) 342-1227

MAINE
Department of Environmental
  Protection
Bureau of Oil and Hazardous
  Materials Control
State House Station #17
Augusta, Maine 04333
(207) 289-2651

MARYLAND
Department of Health and Mental
  Hygiene
Maryland Waste Management
  Administration
Office of Environmental Programs
201 West Preston Street, Room A3
Baltimore, Maryland 21201
(301) 225-5709

MASSACHUSETTS
Department of Environmental
  Quality Engineering
Division of Solid and Hazardous
  Waste
One Winter Street, 5th Floor
Boston, Massachusetts 02108
(617) 292-5589
(617) 292-5851

MICHIGAN
Michigan Department of Natural
  Resources
Hazardous Waste Division
Waste Evaluation Unit
Box 30028
Lansing, Michigan 48909
(517) 373-2730

MINNESOTA
Pollution Control Agency
Solid and Hazardous Waste Division
1935 West County Road, B-2
Roseville, Minnesota 55113
(612) 296-7282

MISSISSIPPI
Department of Natural Resources
Division of Solid and Hazardous
  Waste Management
P.O. Box 10385
Jackson, Mississippi 39209
(601) 961-5062

MISSOURI
Department of Natural Resources
Waste Management Program
P.O. Box 176
Jefferson City, Missouri 65102
(314) 751-3176
Missouri Hotline:
(800) 334-6946

MONTANA
Department of Health and
  Environmental Sciences
Solid and Hazardous Waste Bureau
Cogswell Building, Room B-201
Helena, Montana 59620
(406) 444-2821

NEBRASKA
Department of Environmental
  Control
Hazardous Waste Management
  Section
P.O. Box 94877
State House Station
Lincoln, Nebraska 68509
(402) 471-2186

NEVADA
Division of Environmental Protection
Waste Management Program
Capitol Complex
Carson City, Nevada 89710
(702) 885-4670

NEW HAMPSHIRE
Department of Health and Human
  Services
Division of Public Health Services
Office of Waste Management
Health and Welfare Building
Hazen Drive
Concord, New Hampshire 03301-6527
(603) 271-4608

NEW JERSEY
Department of Environmental
  Protection
Division of Waste Management
32 East Hanover Street, CN-028
Trenton, New Jersey 08625
Hazardous Waste Advisement
  Program: (609) 292-8341

NEW MEXICO
Environmental Improvement
  Division
Ground Water and Hazardous
  Waste Bureau
Hazardous Waste Section
P.O. Box 968
Santa Fe, New Mexico 87504-0968
(505) 827-2922

NEW YORK
Department of Environmental
  Conservation
Bureau of Hazardous Waste
  Operations
50 Wolf Road, Room 209
Albany, New York 12233
(518) 457-0530
SQG Hotline: (800) 631-0666

NORTH CAROLINA
Department of Human Resources
Solid and Hazardous Waste
  Management Branch
P.O. Box 2091
Raleigh, North Carolina 27602
(919) 733-2178

NORTH DAKOTA
Department of Health
Division of Hazardous Waste
  Management and Special Studies
1200 Missouri Avenue
Bismarck, North Dakota 58502-5520
(701) 224-2366

**NORTHERN MARIANA ISLANDS,
COMMONWEALTH OF**
Department of Environmental and
  Health Services
Division of Environmental Quality
P.O. Box 1304
Saipan, Commonwealth of
  Mariana Islands 96950
Overseas call (670) 234-6984

OHIO
Ohio EPA
Division of Solid and Hazardous
  Waste Management
361 East Broad Street
Columbus, Ohio 43266-0558
(614) 466-7220

OKLAHOMA
Waste Management Service
Oklahoma State Department of
  Health
P.O. Box 53551
Oklahoma City, Oklahoma 73152
(405) 271-5338

OREGON
Hazardous and Solid Waste Division
P.O. Box 1760
Portland, Oregon 97207
(503) 229-6534
Toll Free: (800) 452-4011

PENNSYLVANIA
Bureau of Waste Management
Division of Compliance Monitoring
P.O. Box 2063
Harrisburg, Pennsylvania 17120
(717) 787-6239

PUERTO RICO
Environmental Quality Board
P.O. Box 11488
Santurce, Puerto Rico 00910-1488
(809) 723-8184
– or –
EPA Region II
Air and Waste Management Division
26 Federal Plaza
New York, New York 10278
(212) 264-5175

RHODE ISLAND
Department of Environmental
  Management
Division of Air and Hazardous
  Materials
Room 204, Cannon Building
75 Davis Street
Providence, Rhode Island 02908
(401) 277-2797

SOUTH CAROLINA
Department of Health and
  Environmental Control
Bureau of Solid and Hazardous
  Waste Management
2600 Bull Street
Columbia, South Carolina 29201
(803) 734-5200

SOUTH DAKOTA
Department of Water and Natural
  Resources
Office of Air Quality and Solid Waste
Foss Building, Room 217
Pierre, South Dakota 57501
(605) 773-3153

TENNESSEE
Division of Solid Waste Management
Tennessee Department of Public
  Health
701 Broadway
Nashville, Tennessee 37219-5403
(615) 741-3424

TEXAS
Texas Water Commission
Hazardous and Solid Waste Division
Attn: Program Support Section
1700 North Congress
Austin, Texas 78711
(512) 463-7761

UTAH
Department of Health
Bureau of Solid and Hazardous
  Waste Management
P.O. Box 16700
Salt Lake City, Utah 84116-0700
(801) 538-6170

VERMONT
Agency of Environmental
  Conservation
103 South Main Street
Waterbury, Vermont 05676
(802) 244-8702

VIRGIN ISLANDS
Department of Conservation and
  Cultural Affairs
P.O. Box 4399
Charlotte Amalie, St. Thomas
Virgin Islands 00801
(809) 774-3320
   – or –
EPA Region II
Air and Waste Management Division
26 Federal Plaza
New York, New York 10278
(212) 264-5175

VIRGINIA
Department of Health
Division of Solid and Hazardous
  Waste Management
Monroe Building, 11th Floor
101 North 14th Street
Richmond, Virginia 23219
(804) 225-2667
Hazardous Waste Hotline:
(800) 552-2075

WASHINGTON
Department of Ecology
Solid and Hazardous Waste Program
Mail Stop PV-11
Olympia, Washington 98504-8711
(206) 459-6322
In-State: 1-800-633-7585

WEST VIRGINIA
Division of Water Resources
Solid and Hazardous Waste/
  Ground Water Branch
1201 Greenbrier Street
Charleston, West Virginia 25311

WISCONSIN
Department of Natural Resources
Bureau of Solid Waste Management
P.O. Box 7921
Madison, Wisconsin 53707
(608) 266-1327

WYOMING
Department of Environmental Quality
Solid Waste Management Program
122 West 25th Street
Cheyenne, Wyoming 82002
(307) 777-7752
   – or –
EPA Region VIII
Waste Management Division
  (8HWM-ON)
One Denver Place
999 18th Street
Suite 1300
Denver, Colorado 80202-2413
(303) 293-1502

APPENDIX C

Infectious Waste Management Audit

Yes/No/Comments

Infectious Waste Policy

1. Has a policy been established to define infectious waste?
2. Does it conform to your state rules or applicable federal rules?

Personnel

3. Is normal trash being placed in the infectious waste containers?
4. Do staff filling the collection containers know the difference between infectious and noninfectious waste?
5. Are staff trained to separate infectious waste by type and route of disposal?
6. Are new employees trained about infectious waste management?

Collection and Containers

7. Are all types of infectious waste containers clearly identified?
8. Is infectious waste collected from patient areas and laboratories frequently?
9. Do your collection bags tear often?
10. Is their failure rate acceptable?

Spills   <u>**Yes/No/Comments**</u>

11. Have procedures been established to clean up infectious waste spills?
12. Have staff who handle infectious waste been trained for spill response?
13. Have your procedures been followed when you've had an infectious waste spill?
14. Did your procedures work well?

Storage and Processing

15. Is most infectious waste treated on the day it is collected?
16. Is the storage area(s) posted where infectious waste waits for treatment?
17. Is the size of the storage area adequate?
18. Is infectious waste stored in this area longer than two days? If so, is it refrigerated?
19. Does your daily generation rate ever exceed your daily treatment capacity?

Transportation

20. Do you truck infectious waste to a treatment facility?
21. Is the truck equipped for spill response?
22. Are the drivers familiar with the applicable transportation regulations?
23. Do you ever follow the truck?

Treatment

24. Is your infectious waste treated to render it noninfectious?
25. Does your treatment method comply with your state rules or applicable federal rules?
26. Has the effectiveness of your treatment method been tested?
27. Has it been retested?
28. Does it reliably render your waste noninfectious?
29. Have the treatment equipment operators received training about occupational hazards and safety?

Disposal   <u>Yes/No/Comments</u>

30. Are any of your infectious wastes
    landfilled without treatment?
31. Do you use the sanitary sewer for treated
    or untreated infectious wastes?
32. Do you have a copy of your local sanitary
    sewerage authority's rules?
33. Are they being followed?

Your Management Plan

34. Do you have a written infectious waste
    management plan?
35. Has a person been designated to
    supervise your infectious waste
    management?
36. If your treatment equipment breaks down,
    have alternative treatment methods been
    arranged?
37. Do you know your annual costs of
    infectious waste management?
38. Are records kept of the amount of
    infectious waste generated annually at
    your facility?
39. Has it been increasing within the last few
    years?
40. Does your plan account for trends in
    treatment and disposal costs, waste
    volume, and equipment replacement?

Index

Accidents, 188–189, 194–195
Accreditation bodies, 9. *See also* specific
   types
Acetone, 147
Acid rain, 94
Acquired immune deficiency syndrome
   (AIDS), 35, 37, 38, 186, 194, 205
Acute hazardous waste, 151–152
Aerosols, 187
AIDS. *See* Acquired immune deficiency
   syndrome
Air pollution control devices, 102,
   104–105
Alcohols, 147
American Society of Mechanical
   Engineers (ASME), 99
American Society for Testing and
   Materials (ASTM), 24, 25, 37, 99
Ammonium hydroxide, 156
Animal body parts, 41, 162, 173
Antineoplastic drugs, 142–143, 156
Ash, 60, 94, 126
ASME. *See* American Society of
   Mechanical Engineers
ASTM. *See* American Society for
   Testing and Materials
Attenuated vaccines, 40
Autoclaves, 8, 52, 53, 72, 85, 89, 233.
   See also Steam sterilization
   description of, 73–74
   retorts compared to, 74
Autopsy waste, 39

Bacillus
   stearothermophilus, 85–86
   subtilis, 98, 110
BACT. *See* Best available control
   technology
Beach washups, 11–12
Below regulatory concern (BRC), 163

Best available control technology
   (BACT), 100
Biohazard symbol, 133
Biologicals, 40
Bleach, 215–217
Blood, 34–35, 103
Blood-borne diseases, 35, 186, 187. *See*
   also specific types
Blood-borne pathogens, 64
Blood products, 34–35
Body fluids, 103. *See also* specific types
Body parts, 36, 41, 162, 173
Bomb squad, 157
Bottom ash, 60
BRC. *See* Below regulatory concern
Browning Ferris Industries, Inc., 234
Budgeting, 15
Burial of bodies, 36
Burndown, 99
Burnout, 99

Carbon dioxide, 94
Carts for collection of waste, 49, 50
CDC. *See* Centers for Disease Control
Centers for Disease Control (CDC), 26,
   32, 34, 35, 186, 194, 210
CERCLA. *See* Superfund
Certifying bodies, 9. *See also* specific
   types
Chemical indicators, 85
Chemical treatment, 40, 58, 116–120.
   See also Hammermills
   applicability of, 61, 119
   of hazardous waste, 156
   of mixed wastes, 179
   regulation of, 59, 60, 119–120
Chemical waste, 139, 143–147. *See also*
   Hazardous waste; specific types
   contingency plans for, 200–202
   disposal of, 154–157

emergencies involving, 198–202
exposure to, 187
regulation of, 140–142, 147
spills of, 218
Chemotherapeutic drugs, 142–143
CHEMTREC, 210
Chlorine, 103, 216
Chromic acid, 147
Citizen suits, 9
Clean Air Act, 11
Cleaning compounds, 143
Collection of waste, 49–50
Color coding of containers, 133
Commercial treatment firms. *See* Offsite
    treatment
Communication
  of commitment, 13
  in emergencies, 206, 211
  risk, 12
Community emergency planning,
    197–202
Community hazardous material squads,
    208, 211
Community standards, 26
Compaction, 50, 135
Comprehensive Environmental Response,
    Compensation, and Liability Act
    (Superfund), 139–140
Containers, 44–49, 181. *See also*
    Containment
  audit of, 265
  color coding of, 133
  compatibility of with treatment, 52–53
  impermeability of, 45, 47
  injury from lifting and handling, 188
  labeling of, 46, 48, 250, 251
  plastic, 48
  puncture resistance in, 45
  reusable, 54, 250
  rigidity of, 45
  secondary, 54
  selection of, 46
  single-use, 54
  strength of, 47–48
  tamper resistance in, 46
  for transport, 53

Containment, 5, 43. *See also* Containers
  as impediment to direct steam contact
    in sterilization, 80
  of liquids, 48–49
  of low-level radioactive waste, 166, 172
  of sharps, 44–46
  of solid waste, 46–48
  for steam sterilization, 80, 81–82
Contaminated equipment, 41
Contingency planning, 13, 23, 200–202,
    207–214, 237
Controlled air incinerators, 95–96
Cooperative ventures for treatment, 58,
    240
Costs
  control of, 12, 57, 232–233
  of disposal, 130, 232
  immediate, 232
  of incinerator construction, 102–103
  of incinerator operation, 102–103
  of minimization of waste, 233
  of offsite management, 233
  source separation and, 133
  of training for source separation,
    133–134
  of transport, 157
  of treatment, 65–66
Cremation, 36
Cultures, 35–36, 38, 187
Cytotoxic agents, 142–143, 156. *See
    also* specific types

Databases, 210
DCCP. *See* Discarded commercial
    chemical products
Decay in storage (DIS), 166, 167–168
  of mixed wastes, 179, 180
Decisionmaking, 15, 210
Degreasing solvents, 143
De minimis wastes, 163–164
Department of Health and Human
    Services (DHHS), 35, 186
Department of Labor, 186
Deregulated wastes, 163–164
Destruction efficiency, 98–99, 102, 104,
    156, 157

Determination, 5
DHHS. *See* Department of Health and
   Human Services
Diagnostic kits, 38
Difficult wastes, 157. *See also* specific
   types
Dilution, 5, 166. *See also* Disposal
Dioxin, 94, 97
DIS. *See* Decay in storage
Discarded commercial chemical products
   (DCCP), 142
Disease, 37, 185-188. *See also* specific
   types
   blood-borne, 35, 186, 187
   sources of, 185-186
   transmission of, 31, 186
Disinfection, 215-217
   chemical. *See* Chemical treatment
Dispersal, 5, 166. *See also* Disposal
Disposables, 131-132
Disposal, 125-128. *See also* Treatment
   audit of, 267
   of chemical waste, 154-157
   costs of, 130, 232
   defined, 5
   environmental risks of, 8
   of low-level radioactive waste, 170
   offsite. *See* Offsite treatment
   onsite. *See* Onsite treatment
   permits for, 149
   of steam sterilized waste, 89-90
DOL. *See* Department of Labor
Dosimetry badges, 188
Double bagging, 82
Dry heat sterilization. *See* Thermal
   inactivation
Dry waste, 161
Dumping, 8
Dusts, 187

Economic risks, 10
Elementary neutralization, 156
Emergencies, 197-219
   anticipation of, 202
   chemical, 198-202
   communications in, 206, 211

community planning for, 197-202
coordinators of, 210, 214
decisionmaking in, 210
infectious agent, 202
legal requirements in, 212
local authorities in, 211
preparedness for, 206-207
prevention of, 206-207
procedures in, 210-211, 214
radioactive material, 202
regional nuclear, 198
reports after, 219
resource identification in, 211-212
severity of, 204
training for, 219
Emission controls, 60
Engineering controls, 206
Environmental impact, 7, 8, 59, 65,
   130
Equipment. *See also* Incinerators; Steam
   sterilizers; specific types
   breakdown of, 205
   contaminated, 41
   for onsite treatment, 56
   operators of, 190
   permits for, 59
   for steam sterilization, 72-74, 86, 91
Ethylene oxide, 114-115
Evaporation, 169
Exempt wastes, 163-164
Explosives, 157
Exported waste, 253
Extraction procedure toxicity test, 60

Federal guidelines, 25-26
Federal regulation, 20-21. *See also*
      specific laws; specific types
   of chemical waste, 140
   of hazardous waste, 139-140
   of incinerators, 101
   of low-level radioactive waste, 159,
      160
Fire departments, 198, 199
Flyash, 60, 94, 126
Formaldehyde, 114, 115-116
Furans, 94, 97

Gas sterilization, 113–116
Generation, 149
  defined, 5
  MWTA requirements for, 251–253
Generator status, 150–152
Glass, 62, 103
Grinding, 135
Guidelines, 25–26
  defined, 19
  federal, 25–26
  state, 26

Hammermills, 60, 61, 62. *See also*
    Chemical treatment
  appearance of waste after, 63
  costs of, 66
  ease/difficulty of operation of, 60
  load standardization for, 63
  occupational hazards in, 65
  reliability of, 66
  volume reduction from, 64
Handling, 43, 49–51, 177
Hands-on training, 223–224
Hazard analysis, 202–205
Hazardous, defined, 177
Hazardous material, 5. *See also*
    Hazardous waste
  analysis of, 204
  defined, 4–5
  emergencies with. *See* Emergencies
  training in, 214
Hazardous material (HazMat) squads,
    208, 211
Hazardous waste. *See also* Chemical
    waste; Hazardous material; specific
    types
  acute, 151–152
  characteristic, 141
  chemical treatment of, 156
  classification of, 5
  contacts concerning, 259–264
  defined, 4, 140–141, 177
  determination of, 150
  exposure to, 187
  listed, 142
  minimum requirements for, 214

regulation of, 139–140
regulation of institutions generating,
    148–152
Health-care providers, 189
Heat conductivity, 48
Heat exchangers, 59
Heat inactivation. *See* Thermal
    inactivation
Heat-labile plastics, 134
Heat recovery, 57, 153
Heat sterilization. *See* Thermal inactivation
Hemodialysis waste, 39
HEPA. *See* High-efficiency particulate air
Hepatitis B, 22, 35, 37, 186, 205
High-efficiency particulate air (HEPA)
    filters, 115
High-risk waste, 89
HIV. *See* Human immunodeficiency virus
Housekeeping personnel, 189–190
Human blood, 34–35
Human immunodeficiency virus (HIV),
    22, 26, 186, 194
Hydrocarbons, 97
Hydrochloric acid, 156
Hydrogen chloride, 94, 104
Hypochlorite, 179
Hypodermic needles, 36, 37, 205. *See
    also* Sharps

Immediate costs, 232
Immunizations, 193, 207
Incineration, 8, 51, 53, 58, 93–106,
    155. *See also* Incinerators
  advantages of, 93
  air pollution control devices for, 102,
    104–105
  appearance of waste after, 63
  applicability of, 62, 103–104
  automation of, 104
  of biologicals, 40
  of blood, 35
  burndown in, 99
  burnout in, 99
  chemistry of, 97–98
  comparison of with other treatment
    methods, 106

costs of, 65, 102–103
destruction efficiency of, 98–99, 156
disadvantages of, 93–94
emission controls for, 60
engineering of, 103–106
heat recovery from, 57
instrumentation for, 104
load standardization for, 62
of low-level radioactive waste,
    169–170
of mixed wastes, 180
occupational hazards in, 65
onsite. *See* Onsite treatment
operation of, 103–106
regulation of, 60
separation before, 62
of sharps, 37
solid residue from, 80
temperature requirements for, 99
time requirements for, 99
volume reduction from, 64
Incinerators. *See also* Incineration
building of new, 101–103
consultants on, 102
controlled air, 95–96
costs of, 65, 102–103
description of, 95–99
destruction and removal efficiency of,
    102, 104
ease/difficulty of operation of, 60
evaluation of, 101
location of, 106
maintenance of, 205
operators needed for, 61
permits for, 59
public opposition to, 94
regulation of, 100, 101
reliability of, 66
retrofitting of, 101–103
training of operators for, 105–106
trash, 11
Incomplete combustion, 97
Infectious agents, 35–36. *See also*
    Infectious waste; specific types
emergency planning for, 202
spills of, 215–218

Infectious waste. *See also* Infectious
    agents
audit of management of, 265–267
classification of, 32, 34
containers for, 44–49
defined, 5, 31, 133, 134, 265
federal guidelines for management of,
    25–26
handling of, 43, 49–51
identification of, 31–42. *See also*
    specific types
incineration of. *See* Incineration
management of. *See* Management
microbial load of, 32
miscellaneous, 41–42
offsite treatment of. *See* Offsite
    treatment
onsite treatment of. *See* Onsite
    treatment
potential infectious approach to, 33
risks of, 7–10
spills of, 215–218
state guidelines for management of, 26
steam sterilization of. *See* Steam
    sterilization
storage of, 50–51
transport of, 53–54
treatment of. *See* specific types;
    Treatment
types of, 34–42
Infectious waste stream, 62
Injuries, 38, 188–189
Iodine, 179
Irradiation, 58, 120

JCAHO. *See* Joint Commission on the
    Accreditation of Healthcare
    Organizations
Joint Commission on the Accreditation
    of Healthcare Organizations
    (JCAHO), 24–25, 32, 188, 222

Labeling of containers, 46, 48, 250, 251
Laboratory waste, 35, 38–39, 143,
    144–147, 156. *See also* specific types
regulation of, 147

Landfills, 8, 11, 60, 125–128, 130, 156
  capacity of, 129
  chemical waste and, 154
  tipping fees for, 63, 64
  workers at, 190–191
Lawsuits, 9
Legal liabilities, 8–10
LEPC. *See* Local Emergency Planning
  Committee
Liabilities, 7, 8–10, 12, 237
Liquid scintillation cocktails (LSC),
  161–162, 172–173
Liquid waste
  chemical disinfection of, 119
  containment of, 48–49
  incineration of, 103
  low-level radioactive, 162, 174
  sewer disposal of, 121, 128
  steam sterilization of, 48
  thermal treatment of, 61
Litigation, 9
Live vaccines, 40
Load standardization, 62–63
Local Emergency Planning Committee
  (LEPC), 198, 199, 212
Local fire departments, 198, 199
Local regulation, 23–24
  of hazardous waste, 139–140
  of incinerators, 101
  of low-level radioactive waste, 159
Low-level radioactive waste, 159–174.
    See also specific types
  composition of, 160–162
  concentration of, 166
  containment of, 166, 172
  defined, 160
  de minimis, 163–164
  deregulated, 163–164
  dilution of, 166
  dispersal of, 166
  disposal of, 170
  exempt, 163–164
  fuel use of, 169–170
  incineration of, 169–170
  indefinite storage of, 171
  liquid, 162, 174

  minimization of, 159, 170–171
  offsite management of, 171–172
  onsite management of, 167–171
  performance-based standards for,
    164–165
  planning for management of, 159,
    165–167
  regulation of, 159, 160, 162–165, 166
  sewer disposal of, 168–169, 174
  shallow land burial of, 171–172
  storage of, 166, 167–168, 171
  strategy for management of, 166–167
Low-Level Radioactive Waste Policy Act
    of 1980, 160, 164
LSC. *See* Liquid scintillation cocktails

Maintenance activities, 143, 190
Management, 6, 265–267. *See also*
    Treatment
  audit of current practices in, 231–232
  cost containment and, 232–233
  essential components of, 231–240
  federal regulation of. *See* Federal
    regulation
  flexible strategies for, 16
  implementation of plan for, 15–16
  of mixed wastes, 178, 180–181
  national trends in, 10–11
  objectives of, 12–13
  offsite. *See* Offsite treatment
  onsite. *See* Onsite treatment
  planning for, 13–15
  policies in, 15
  procedures in, 15–16
  responsibility assignment in, 16
  training in, 15–16
Marking of containers, 46, 48, 250, 251
Medical surveillance, 206–207
Medical waste, defined, 5
Medical Waste Tracking Act of 1988
    (MWTA), 11, 20–21, 100, 189,
    245–256
  demonstration program of, 256
  generator requirements of, 251–253
  labeling requirements of, 250
  mixed wastes and, 178

onsite incineration requirements of, 251–253
pretransport requirements of, 249–251
purpose of, 245
recordkeeping requirements of, 254, 255
reporting requirements of, 254–255
scope of, 249
separation requirements of, 249–250
storage requirements of, 250
transport requirements of, 253–255
types of waste regulated under, 245–248
Mercury, 94, 154
Metals, 94, 156, 157. See also specific types
Microbial load, 32
Mineral acids, 156
Minimization of surplus materials, 153
Minimization of waste, 7, 11, 94, 129–135, 153
benefits of, 130
as cost containment, 232–233
costs of, 233
defined, 139
disposables and, 131–132
low-level radioactive waste and, 159, 170–171
mixed wastes and, 180
reasons for, 129–131
source separation and, 132–134
Mixed wastes, 61, 165, 177–181
handling of, 177
management of, 178, 180–181
minimization of, 180
priority to greatest risk in, 179
reduction in quantities of, 181
regulation of, 178–179
sources of, 177–178, 181
storage of, 179, 180, 181
treatment of, 177, 180
Morticians, 36
Multiple-hazard wastes. See Mixed wastes
MWTA. See Medical Waste Tracking Act

National Cancer Institute, 215
National Committee for Clinical Laboratory Standards, 26
National Institute for Occupational Safety and Health (NIOSH), 26
National Institutes of Health (NIH), 35
National Response Team, 212
National trends in waste management, 10–11
Needles, 36, 37, 205. See also Sharps
Neutralization, 156
NIH. See National Institutes of Health
NIMBY (Not In My Back Yard), 94
NIOSH. See National Institute for Occupational Safety and Health
Nitric acid, 156
Nitrogen oxides, 94
Not In My Back Yard (NIMBY), 94
NRC. See Nuclear Regulatory Commission
Nuclear Regulatory Commission (NRC), 21, 162, 164, 166–167, 170
emergencies and, 202
licenses from, 163
mixed wastes and, 178, 180
offsite facility licensing by, 172
radiation exposure regulation of, 188

Occupational risks, 7–8, 64–65, 130, 185–195. See also specific types
of disease, 185–188. See also Disease
of equipment operators, 190
of health-care providers, 189
of housekeeping personnel, 189–190
of injury, 38, 188–189
job training and, 193–194
of landfill workers, 190–191
of maintenance workers, 190
minimization of waste as reduction in, 130
reduction in, 130, 191–195
of trash haulers, 190
Occupational Safety and Health Administration (OSHA), 22, 23, 26, 32, 37, 41
blood-borne pathogens and, 64
formaldehyde and, 116

sharps and, 45, 46, 189
training and, 222, 226
Offsite treatment, 57–58, 156–157,
    233–239
  bid specifications for, 238–239
  business of, 234–235
  cost of, 233
  disadvantages of, 57–58
  evaluation of facilities for, 58,
    236–238
  of low-level radioactive waste,
    171–172
  requests for proposals for, 238–239
  risks of, 235–236
  transport for, 53–54
Onsite treatment, 56–57, 155, 233–234
  advantages of, 57
  disadvantages of, 57
  of low-level radioactive waste,
    167–171
  MWTA requirements for, 252–253
  physical constraints against, 56
Open dumping, 8
Organic solvents, 147, 172. See also
    specific types
OSHA. See Occupational Safety and
    Health Administration
Oxidation, 99

PAHs. See Polyaromatic hydrocarbons
Paints, 143
Partial oxidation, 97
Pathological waste, 36. See also specific
    types
  disposal of, 125, 127
  incineration of, 155
Patient care waste, 39–40
Permits, 59, 149, 157
  chemical treatment, 156
  incinerator, 96, 155
  limited burn, 180
Personal dosimetry badges, 188
Personal protective equipment, 41
Petroleum distillates, 147
Photography laboratories, 153
Picric acid, 147

PICs. See Products of incomplete
    combustion
Planning, 13
  audit of, 267
  community emergency, 197–202
  contingency, 13, 200–202, 207–214,
    233, 237
  for low-level radioactive waste
    management, 159, 165–167
  process of, 207–209
  for waste management, 13–15
Plastic bags, 46, 48, 52, 53, 127–128
  in steam sterilization, 81–82
Plastic containers, 48
Plastics, 62. See also specific types
  heat-labile, 134
  incineration of, 103–104
Political risks, 10, 159
Pollution control devices, 102, 104–105
Polyaromatic hydrocarbons (PAHs), 97
Polyaromatic organic matter (POM), 97
Polychlorinated dibenzo-p-dioxins
    (dioxin), 94, 97
Polychlorinated dibenzofurans (furans),
    94, 97
Polyethylene, 81
Polypropylene, 81
Polyvinyl chloride (PVC), 53
POM. See Polyaromatic organic matter
Potential infectiousness, 33
Pretransport requirements under MWTA,
    249–251
Prevention of waste. See Minimization
    of waste
Preventive medicine, 206–207
Procedural controls, 206
Products of incomplete combustion
    (PICs), 97
Product substitution, 153
Professional organizations, 19, 24. See
    also specific organizations
Pseudocumene, 172
Public Health Service, 35
Public opposition to incinerators, 94
Public perception of risks, 10
Puncture resistance in containers, 45

PVC. *See* Polyvinyl chloride
Pyrolysis, 97

QA/QC. *See* Quality assurance/quality
    control
Quality assurance/quality control
    (QA/QC), 55, 68–70, 121
  for chemical disinfection, 120
  for heat inactivation, 113
  for steam sterilization, 90–91

Radioactive material. *See also*
    Radioactive waste; specific types
  spills of, 218–219
Radioactive waste, 179. *See also* specific
    types
  emergency planning for, 202
  exposure to, 187–188
  low-level. *See* Low-level radioactive
    waste
  spills of, 218–219
Radiolabeled iodine, 179
Radionuclides, 159, 160, 161, 173. *See*
    also Low-level radioactive waste
  decay of, 168
  incineration of, 180
  liquid, 174
  short-half-life, 181
RCRA. *See* Resource Conservation and
    Recovery Act
Reactive chemicals, 157
Reagents, 147
Recycling, 8, 11, 153, 157
"Red bags," 48, 52, 127–128
Reduction of volume, 63–64
Reduction of waste, 7. *See also*
    Minimization of waste
Regional nuclear disasters, 198
Regional treatment facilities, 58, 240
Regulation, 19–24. *See also* specific
    laws; specific types
  of chemical disinfection, 119–120
  of chemical waste, 140–142, 147
  defined, 19
  of emergency procedures, 212
  federal. *See* Federal regulation

  of incinerators, 100, 101
  of institutions generating hazardous
    waste, 148–152
  local.. *See* Local regulation
  of low-level radioactive waste, 159,
    160, 162–165, 166
  of mixed wastes, 178–179
  of onsite treatment, 56
  right-to-know, 221, 222–223
  state. *See* State regulation
  of training, 221–222
  of treatment, 59–60
Requests for proposals (RFPs), 238–239
Resource Conservation and Recovery
    Act of 1976 (RCRA), 11, 20, 21,
    138–140
  incinerators and, 96, 102, 104
  landfills and, 60
  mixed wastes and, 178, 180
  spills and, 218
Response
  to accidents, 194–195
  to emergencies. *See* Emergencies
  to exposures, 194
  to spills, 214–219
Responsibility assignment, 16
Retorts, 72, 74, 85. *See also* Steam
    sterilization
Reusables, 131, 250
Reuse, 153
RFPs. *See* Requests for proposals
Right-to-know regulation, 221, 222–223
Risk analysis, 202, 204, 209
Risks, 7–10. *See also* specific types
  communication of, 12
  economic, 10
  environmental, 7, 8, 130
  of gas sterilization, 114–116
  measurement of, 7
  minimization of waste as reduction in,
    130, 133
  occupational. *See* Occupational risks
  political, 10, 159
  public perception of, 10
  reduction in, 12
  social, 10

SARA. *See* Superfund Amendments and
    Reauthorization Act
Scrubber discharges, 94
Secondary containers, 54
Segregation of wastes. *See* Separation of
    wastes
Semiliquids, 128
Separation of wastes, 132–134, 165, 233
    importance of, 62
    MWTA requirements for, 249–250
Sewer system, 120–121, 128, 154–155,
    156
    low-level radioactive waste and,
        168–169, 174
Shallow land burial, 171–172
Sharps, 36–38, 205. *See also* specific
    types
    beach washups of, 11–12
    containment of, 44–46
    disposal of, 125, 126–127
    handling of, 43
    heat inactivation of, 111
    injury from, 189
    in landfills, 191
    puncture resistance in, 45
    rigidity of, 45
Shipment. *See* Transport
Shredding, 135
Silver recovery, 153
Slags, 103
Social risks, 10
Sodium hypochlorite, 216
Solid waste. *See also* specific types
    chemical disinfection of, 119
    containment of, 46–48
    defined, 4
Solvents, 143, 147, 153, 157, 172. *See
    also* specific types
Source separation, 132–134, 165
Specimens, 38, 187
Spill kits, 215
Spills, 214–219, 266
Stack emissions from incineration, 94
Standards, 24–26. *See also* specific types
    ASME, 99
    ASTM, 25, 99

community, 26
defined, 19
JCAHO, 24–25
load, 62–63
for low-level radioactive waste,
    164–165
wastewater, 60
State Emergency Response Commission,
    198, 199
State guidelines, 26
State regulation, 23, 258
    of chemical waste, 140
    of hazardous waste, 139–140
    of incinerators, 100, 101
    of low-level radioactive waste, 159,
        162
Steam sterilization, 8, 51–52, 58, 71–91.
    See also Autoclaves; Retorts; Steam
    sterilizers
    alternative conditions for, 83–84
    appearance of waste after, 63
    applicability of, 61, 62, 80, 81
    of blood, 35
    of chemical waste, 154
    conditions required for, 71–76
    containment for, 80, 81–82
    control of, 75–76
    costs of, 65–66
    of cultures, 35
    depth in, 83
    direct steam contact in, 76, 78–80
    disadvantages of, 89
    disposal of waste after, 89–90
    ease/difficulty of operation of, 60
    effectiveness of, 61
    equipment for, 86, 91. *See also* Steam
        sterilizers
    excessive mass or weight in, 77–78
    exposure period in, 75, 83
    heat capacity in, 78
    heat conductivity in, 78
    heat transfer barriers in, 78
    of high-risk waste, 89
    instrumentation in, 75–76, 91
    interference with heating in, 77–78
    of laboratory waste, 35

labor requirements for, 88
of liquid waste, 48
load carriers in, 81–82
load configuration in, 81–82
load standardization in, 62
of mixed wastes, 180
occupational hazards in, 64
operator training for, 88–89
optimal conditions for, 80–84
physical requirements for, 87–88
potential problems in, 76–80
in practice, 87–89
quality assurance and control in,
   90–91
recordkeeping in, 89
regulation of, 59
requirements for effectiveness in, 52
role of steam in, 72
saturated steam in, 72
separation before, 62
of sharps, 37
standard conditions for, 86
standard load for, 84–85
standard operating procedures for,
   84–87, 91
steam production in, 72
steam quality in, 80
temperature requirements for, 71, 76,
   83–84
time requirements for, 71, 76
vacuum cycles in, 83
volume reduction in, 64, 134
Steam sterilizers. See also Steam
   sterilization
contamination of by mixed wastes, 179
costs of, 65–66
features of, 72–74
maintenance of, 205
operation of, 74–75
reliability of, 66
training of operators of, 88–89
types of, 72–74
Sterilization. See also specific types
of chemical waste, 154
defined, 85
gas, 113–116

heat. See Thermal inactivation
steam. See Steam sterilization
Stocks of infectious agents, 35–36
Storage, 50–51, 266
for decay. See Decay in storage (DIS)
defined, 5
of low-level radioactive waste, 166,
   167–168, 171
of mixed wastes, 179, 180, 181
MWTA requirements for, 250
permits for, 149
Strategic objectives, 12–13
Substitution of materials, 181
Sulfuric acid, 156
Superfund, 139–140
Superfund Amendments and
   Reauthorization Act (SARA), 140,
   198–200
Surgical waste, 39
Surplus materials. See also specific types
minimization of, 153
return of to manufacturer, 156
Syringes, 11–12, 36, 37. See also Sharps

Tamper resistance of containers, 46
Temperature
for chemical disinfection, 118
for incineration, 99
for steam sterilization, 71, 76, 83–84
Thermal inactivation, 40, 58, 59,
   109–113
applicability of, 110–112
ease/difficulty of operation of, 60
of liquid waste, 61
Thermocouples, 76, 85
Thinners, 143
Toluene, 172, 173
Toxic metals, 94. See also specific types
Training, 265
course content in, 223
for emergencies, 219
follow-up to, 225–226
hands-on, 223–224
hazardous material, 214
of incinerator operators, 105–106
institutional policies on, 221

instructors for, 227–228
JCAHO requirements for, 222
job, 193–194
program for, 222–228
recordkeeping for, 228
regulation of, 221–222
schedule of, 226
for source separation, 133–134
of staff, 221–228
of steam sterilizer operators, 88–89
testing after, 225
of waste handlers, 221–228
in waste management, 15–16
Transport, 53–54
audit of, 266
containers for, 53
costs of, 157
MWTA requirements for, 253–255
permits for, 149
Trash haulers, 190
Trash incinerators, 11
Treatment, 7, 8, 51–53, 55–70. *See also*
    specific types
applicability of, 61–62
audit of, 266
comparison of, 67–68
compatibility of containers with, 52–53
cooperative ventures for, 58, 240
costs of, 65–66
defined, 5
ease/difficulty of operation of, 60
effect of on waste, 63
environmental effects of, 65
of mixed wastes, 177, 180

occupational hazards in, 64–65
offsite. *See* Offsite treatment
onsite. *See* Onsite treatment
permits for, 149
regional facilities for, 58, 240
regulation of, 59–60
reliability of, 6
selection of, 58–67
selection of place for, 55, 56–58
skilled operator for, 61
types of, 55. *See also* specific types
University of Minnesota, 130, 131–132,
    133
University of North Carolina, 129

Vaccines, 40
Volume reduction, 63–64, 134–135

Waste generation. *See* Generation
Waste management. *See* Management
Waste Management, Inc., 234
Waste treatment. *See* Treatment
Wastewater standards, 60
Wastewater treatment, 58, 111
Water treatment chemicals, 143
Weight reduction, 134–135
Wet scrubbers, 60
Wet waste, 187

X-ray facilities, 153
Xylene, 172, 173

Z third criterion, 71